ALGORITHMIC AND REGISTER-TRANSFER LEVEL SYNTHESIS:
THE SYSTEM ARCHITECT'S WORKBENCH

THE KLUWER INTERNATIONAL SERIES IN ENGINEERING AND COMPUTER SCIENCE

VLSI, COMPUTER ARCHITECTURE AND DIGITAL SIGNAL PROCESSING

Consulting Editor
Jonathan Allen

Other books in the series:

Logic Minimization Algorithms for VLSI Synthesis. R.K. Brayton, G.D. Hachtel, C.T. McMullen, and A.L. Sangiovanni-Vincentelli. ISBN 0-89838-164-9.
Adaptive Filters: Structures, Algorithms, and Applications. M.L. Honig and D.G. Messerschmitt. ISBN 0-89838-163-0.
Introduction to VLSI Silicon Devices: Physics, Technology and Characterization. B. El-Kareh and R.J. Bombard. ISBN 0-89838-210-6.
Latchup in CMOS Technology: The Problem and Its Cure. R.R. Troutman. ISBN 0-89838-215-7.
Digital CMOS Circuit Design. M. Annaratone. ISBN 0-89838-224-6.
The Bounding Approach to VLSI Circuit Simulation. C.A. Zukowski. ISBN 0-89838-176-2.
Multi-Level Simulation for VLSI Design. D.D. Hill and D.R. Coelho. ISBN 0-89838-184-3.
Relaxation Techniques for the Simulation of VLSI Circuits. J. White and A. Sangiovanni-Vincentelli. ISBN 0-89838-186-X.
VLSI CAD Tools and Applications. W. Fichtner and M. Morf, editors. ISBN 0-89838-193-2.
A VLSI Architecture for Concurrent Data Structures. W.J. Dally. ISBN 0-89838-235-1.
Yield Simulation for Integrated Circuits. D.M.H. Walker. ISBN 0-89838-244-0.
VLSI Specification, Verification and Synthesis. G. Birtwistle and P.A. Subrahmanyam. ISBN 0-89838-246-7.
Fundamentals of Computer-Aided Circuit Simulation. W.J. McCalla. ISBN 0-89838-248-3.
Serial Data Computation. S.G. Smith and P.B. Denyer. ISBN 0-89838-253-X.
Phonologic Parsing in Speech Recognition. K.W. Church. ISBN 0-89838-250-5.
Simulated Annealing for VLSI Design. D.F. Wong, H.W. Leong, and C.L. Liu. ISBN 0-89838-256-4.
Polycrystalline Silicon for Integrated Circuit Applications. T. Kamins. ISBN 0-89838-259-9.
FET Modeling for Circuit Simulation. D. Divekar. ISBN 0-89838-264-5.
VLSI Placement and Global Routing Using Simulated Annealing. C. Sechen. ISBN 0-89838-281-5.
Adaptive Filters and Equalizers. B. Mulgrew, C.F.N. Cowan. ISBN 0-89838-285-8.
Computer-Aided Design and VLSI Device Development, Second Edition. K.M. Cham, S-Y. Oh, J.L. Moll, K. Lee, P. Vande Voorde, D. Chin. ISBN: 0-89838-277-7.
Automatic Speech Recognition. K-F. Lee. ISBN 0-89838-296-3.
Speech Time-Frequency Representations. M.D. Riley. ISBN 0-89838-298-X
A Systolic Array Optimizing Compiler. M.S. Lam. ISBN: 0-89838-300-5.
Algorithms and Techniques for VLSI Layout Synthesis. D. Hill, D. Shugard, J. Fishburn, K. Keutzer. ISBN: 0-89838-301-3.
Switch-Level Timing Simulation of MOS VLSI Circuits. V.B. Rao, D.V. Overhauser, T.N. Trick, I.N. Hajj. ISBN 0-89838-302-1
VLSI for Artificial Intelligence. J.G. Delgado-Frias, W.R. Moore (Editors). ISBN 0-7923-9000-8.
Wafer Level Integrated Systems: Implementation Issues. S.K. Tewksbury. ISBN 0-7923-9006-7
The Annealing Algorithm. R.H.J.M. Otten & L.P.P.P. van Ginneken. ISBN 0-7923-9022-9.
VHDL: Hardware Description and Design. R. Lipsett, C. Schaefer and C. Ussery. ISBN 0-7923-9030-X.
The VHDL Handbook. D. Coelho. ISBN 0-7923-9031-8.
Unified Methods for VLSI Simulation and Test Generation. K.T. Cheng and V.D. Agrawal. ISBN 0-7923-9025-3
ASIC System Design with VHDL: A Paradigm. S.S. Leung and M.A. Shanblatt. ISBN 0-7923-9032-6.
BiCMOS Technology and Applications. A.R. Alvarez (Editor). ISBN 0-7923-9033-4.
Analog VLSI Implementation of Neural Systems. C. Mead and M. Ismail (Editors). ISBN 0-7923-9040-7.
The MIPS-X RISC Microprocessor. P. Chow. ISBN 0-7923-9045-8.
Nonlinear Digital Filters: Principles and Applications. I. Pitas and A.N. Venetsanopoulos. ISBN 0-7923-9049-0.

ALGORITHMIC AND REGISTER-TRANSFER LEVEL SYNTHESIS: THE SYSTEM ARCHITECT'S WORKBENCH

by

D.E. Thomas
E.D. Lagnese
R.A. Walker
J.A. Nestor
J.V. Rajan
R.L. Blackburn

Carnegie Mellon University

KLUWER ACADEMIC PUBLISHERS
Boston/Dordrecht/London

Distributors for North America:
Kluwer Academic Publishers
101 Philip Drive
Assinippi Park
Norwell, Massachusetts 02061 USA

Distributors for all other countries:
Kluwer Academic Publishers Group
Distribution Centre
Post Office Box 322
3300 AH Dordrecht, THE NETHERLANDS

Library of Congress Cataloging-in-Publication Data

Algorithmic and register-transfer level synthesis : the system architect's workbench / by D.E. Thomas ... [et al.].
p. cm. — (The Kluwer international series in engineering and computer science. VLSI, computer architecture, and digital signal processing)
Includes bibliographical references.
ISBN 0-7923-9053-9
1. Integrated circuits—Design and construction. 2. Computer architecture. I. Thomas, D. E. (Donald E.), 1951- . II. Series.
TK7874.A418 1990
621.39'5—dc20 89-28031
CIP

Printed in the United States of America

to our spouses

and to Robert B.'s many friends

Table of Contents

Acknowledgements

We would like the thank the many people who have been associated with the System Architect's Workbench project, and its prececessor systems, over the years, including the following people: Stephen Director, Daniel Siewiorek, Mario Barbacci, Alice Parker, John Grason, Edward A. Snow, Michael C. McFarland, S.J., David A. Gatenby, John F. Lertola, Lou Hafer, C.Y. Hitchcock III, Thaddeus J. Kowalski, David J. Geiger, Patti M. Koenig, Edward D. Un, Rajeev Murgai, David L. Springer, Richard Cloutier, Anthony Gadient, John Hagerman, Brian Ebert, John Eppling, Rajen Ramchandani, Kees Vissers, and Michel Doreau.

This work has been supported by the Semiconductor Research Corp., Digital Equipment Corporation, General Electric, Gateway Design Automation, General Motors, GTE, Hewlett Packard, Intel, IBM, ITT, Philips, Raytheon, Texas Instruments, the Army Research Office, and the National Science Foundation.

ALGORITHMIC AND REGISTER-TRANSFER LEVEL SYNTHESIS: THE SYSTEM ARCHITECT'S WORKBENCH

1. Introduction

Recently there has been increased interest in the development of computer-aided design programs to support the system level designer of integrated circuits more actively. Such design tools hold the promise of raising the level of abstraction at which an integrated circuit is designed, thus releasing the current designers from many of the details of logic and circuit level design. The promise further suggests that a whole new group of designers in neighboring engineering and science disciplines, with far less understanding of integrated circuit design, will also be able to increase their productivity and the functionality of the systems they design. This promise has been made repeatedly as each new higher level of computer-aided design tool is introduced and has repeatedly fallen short of fulfillment. This book presents the results of research aimed at introducing yet higher levels of design tools that will inch the integrated circuit design community closer to the fulfillment of that promise.

1.1. SYNTHESIS OF INTEGRATED CIRCUITS

In the integrated circuit (IC) design process, a behavior that meets certain specifications is conceived for a system, the behavior is used to produce a design in terms of a set of structural logic elements, and these logic elements are mapped onto physical units. The design process is impacted by a set of constraints as well as technological information (i.e. the logic elements and physical units used for the design). Indeed, design of such systems is not just a top-down process. Rather, there must be a close interaction between the design process, the constraints, and feedback on how the design is progressing.

The IC design process is made up of several design steps, many of them exhibiting these characteristics. One well understood step is logic synthesis. In logic synthesis, the behavioral description may be a set of Boolean equations or a truth table. The logic synthesis process, through minimization and translation to a specific cell library, maps the behavior onto structural logic elements. Later design steps consider the physical placement and routing of the design.

This book presents research aimed at providing computer-aided design tools for two higher level steps in the design process: Algorithmic Level synthesis and Register-Transfer Level synthesis.

1.1.1 Algorithmic Level Synthesis

In *Algorithmic Level synthesis*, a behavioral description specifies an algorithm or possibly a set of concurrent algorithms that are to be implemented. The Algorithmic Level design process then considers the partitioning of the behavior on the system's proposed physical units (for instance IC's or boards). Based on analyses of this partitioning, the behavior may be transformed further or altered by adding or deleting functionality.

This level of design is analogous to writing a set of concurrent programs that will be executed on a multiprocessor. Each program is written to run on a different processor (assuming a static assignment of programs to processors) and to interact as necessary with other programs on other processors. Similarly, Algorithmic Level design must deal with the implementation of multiple algorithms, each with their separate threads of control. But in contrast to concurrent programming on a multiprocessor, Algorithmic Level synthesis considers the structural and physical design of each of the "programs" as a separate, specially designed data path and controller.

An illustration of the integrated circuit design process is shown in Figure 1-1. At the top, a behavioral description of the algorithm to be

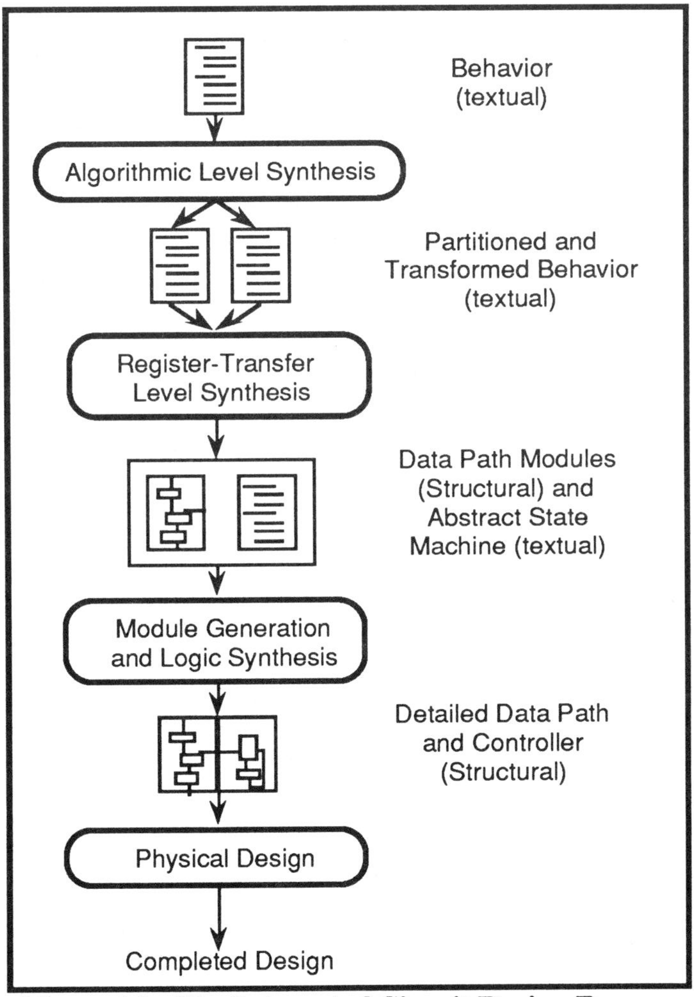

Figure 1-1. The Integrated Circuit Design Process

implemented is conceived by a designer. Through analysis, the designer may determine that the behavior must be altered. Consider two of the possible scenarios that may arise in Algorithmic Level synthesis:

1) The designer may find that the behavior is too large to fit onto a single physical unit, such as an integrated circuit, and thus the behavior must be partitioned across several physical units.

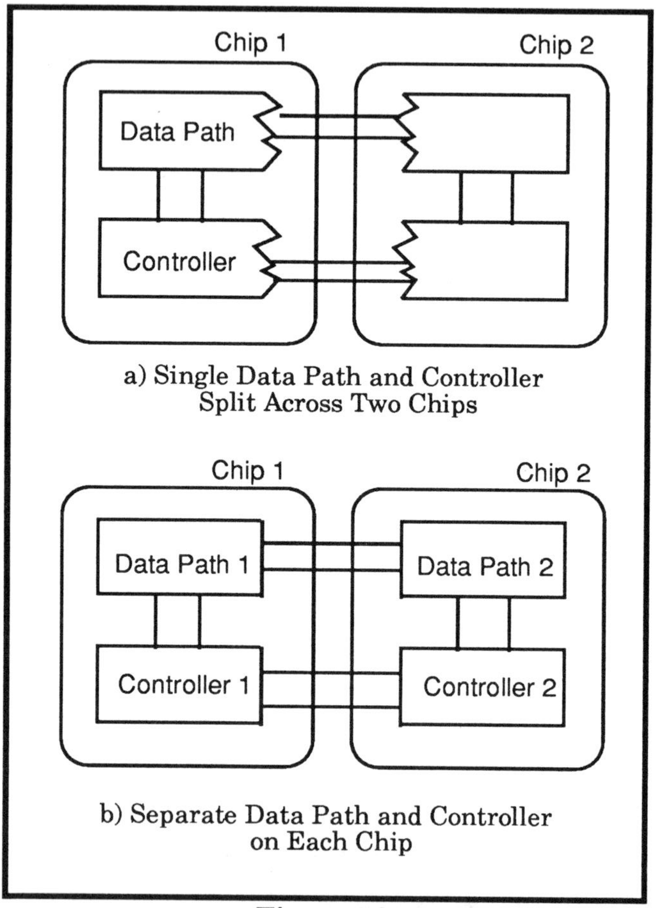

Figure 1-2
Scenario 1: Splitting a Design Across Two Chips

2) The designer may find that certain parts of the behavior interact with other concurrent behaviors and thus have data rate constraints not attainable by the behavior as a whole.

In each of these scenarios, the designer may need to transform the behavior as illustrated in Figure 1-2 in response to physical constraints. In scenario 1, the constraints are area and time. If the design is too large for a single integrated circuit, an implementation of the behavior as it is will require signal paths to connect between the IC's

(See Figure 1-2a). This requirement implies a slower clock to allow for the extra off-chip delays during every clock cycle. An alternate implementation, shown in Figure 1-2b, splits the behavior into two concurrent behaviors, putting each behavior on a separate chip. With this approach, both behaviors would have all of their logic on a single chip and the clock rates for each could be faster than before. Thus the design is only slowed by a few off-chip delays when the two behaviors interact. This partitioning of the behavior is appropriate only when there are few interaction points between the concurrent behaviors.

A second scenario, where the constraint is time, is shown in Figure 1-3. Here the designer finds that the instantaneous rate at which data is being produced at the behavior's input may exceed the instantaneous rate at which it can consume the data. The standard solution to this producer - consumer problem suggests that part of the consumer behavior be split off and coupled with a queue to buffer the peak data rate until the rest of the consumer behavior can consume it. As illustrated in Figure 1-3, what was originally one behavior has now become two concurrent behaviors. Now the producer (not shown) feeds data to the queue behavior which buffers it for consumption. The queue reacts to the bursty arrival rates and allows the consumer to work at its own rate.

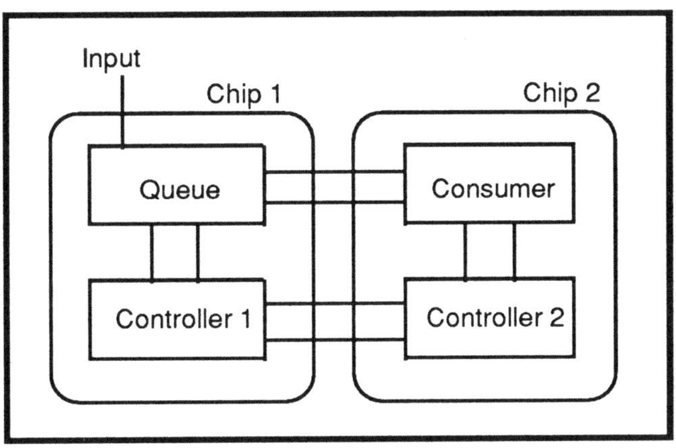

Figure 1-3
Scenario 2: Splitting a Design to Meet Time Constraints

In each of these scenarios, the designer must transform the original behavior due to the physical constraints of implementation. The results of Algorithmic Level synthesis are new behaviors, with pieces of the original behavior extracted and reformed to better meet the physical constraints.

1.1.2. Register-Transfer Level Synthesis

The next step in Figure 1-1 shows a behavioral description being carried through to a Register-Transfer (RT) Level of implementation. In this phase, called *Register-Transfer Level synthesis*, (McFarland [McFarland88] refers to it as high level synthesis), a data path and controller are designed for each of the behaviors. Specifically, four major decisions are made:

- The number of control steps (or control states) needed in the design is determined and the operators specified in the behavior are scheduled into them.

- The number of registers needed in the design is determined and the variables and intermediate values described in the behavior are assigned to them.

- The number of functional units (arithmetic-logic units, adders, etc.) needed in the design is determined and the operators specified in the behavior are assigned to them.

- The number of wires, multiplexors and busses needed in the design is determined and the interconnections between the controller, registers, memories, and operators are assigned to them.

As a result of Register-Transfer Level synthesis, the design is specified in terms of data path modules such as functional units, registers, and interconnections, and a controller module specifying how the data path modules are to be evoked to execute the original behavior.

The data path modules are specified abstractly. For instance, an arithmetic-logic unit (ALU) would have its width and functions (e.g. add, subtract, and OR) specified. As illustrated in Figure 1-1, the controller is described textually as an abstract state machine (states, next states, and data path elements to enable). The controller is not any more detailed since the actual implementations of the data path modules, nor how to enable them, is known at this point in the design process.

Although the above register-transfer design decisions are listed individually, there are many interactions and interdependencies between them. For instance, if the control schedule is determined first, it may specify a level of parallelism that is unimplementable. This might happen if two multiply operators are scheduled to occur in the same control step. Given that two hardware multipliers would be needed for the implementation, they both might not fit on the IC. Making all of the register-transfer decisions at the same time has shown to be far too complex.

The next step in Figure 1-1 shows the data path elements being implemented using module generators and the controller being implemented through state assignment and logic synthesis. At this point, the data path consists of modules such as registers, arithmetic-logic units, and multiplexors. Module generators detail each of these modules by specifying the necessary internal logic structure. Once the internal structure is known, a controller is synthesized to evoke the operations in the data path modules. Control steps are implemented as states in the controller and logic is synthesized to control the data path and select the next state. Finally, the detailed data path and controller modules are placed and routed by a physical design system and the completed design is produced.

1.2 THE SYSTEM ARCHITECT'S WORKBENCH

This book presents the results of several coordinated research projects, collectively referred to as the System Architect's Workbench (*the Workbench*). These research projects were Ph.D. Theses completed under the direction of Professor Donald Thomas and granted by the Electrical and Computer Engineering Department of Carnegie Mellon University. The research described here fits into the first two design steps of Figure 1-1 and can be grouped into three major categories:

- Computer-aided design tools to support Algorithmic Level synthesis.

- Computer-aided design tools to support Register-Transfer Level synthesis.

- Computer-aided design tools to support the visualization of an automatically designed system in terms of its original behavioral description.

The goal of the System Architect's Workbench project is to do research in the areas of Algorithmic and Register-Transfer Level synthesis, and to develop demonstration design tools to support the research. A diagram of the Workbench is shown in Figure 1-4.

The design process starts from the top of the figure and flows downward. A behavioral description enters the system and is translated into an internal format called the *Value Trace* (VT). The tools within the Workbench box work from the VT, altering it, adding information to it, or adding new design information in other data structures. The result of the design tools is a Register-Transfer Level data path and abstract controller as described previously. This output is suitable input for module generators, logic synthesis tools, and physical design tools.

Within the Workbench, there are two separate synthesis paths. On the left is the "general design" path. These tools are aimed at designing a wide range of systems and thus encompass general design information. On the right is the "style-specific" path. This tool is aimed at designing microprocessors, and thus its design algorithms are tuned to the design decisions made for that style of implementation.

Although each individual design tool will be discussed in greater depth later in the book, their functions are discussed here briefly:

- The Mill ISPS-to-VT translator. This program produces the value trace description from the ISPS description of the system to be built.

- The Behavioral Transformations tool provides a means to transform the original behavioral description. These transformations make the behavior more appropriate for implementation and provide the basis for exploring the design space at the Algorithmic Level of design.

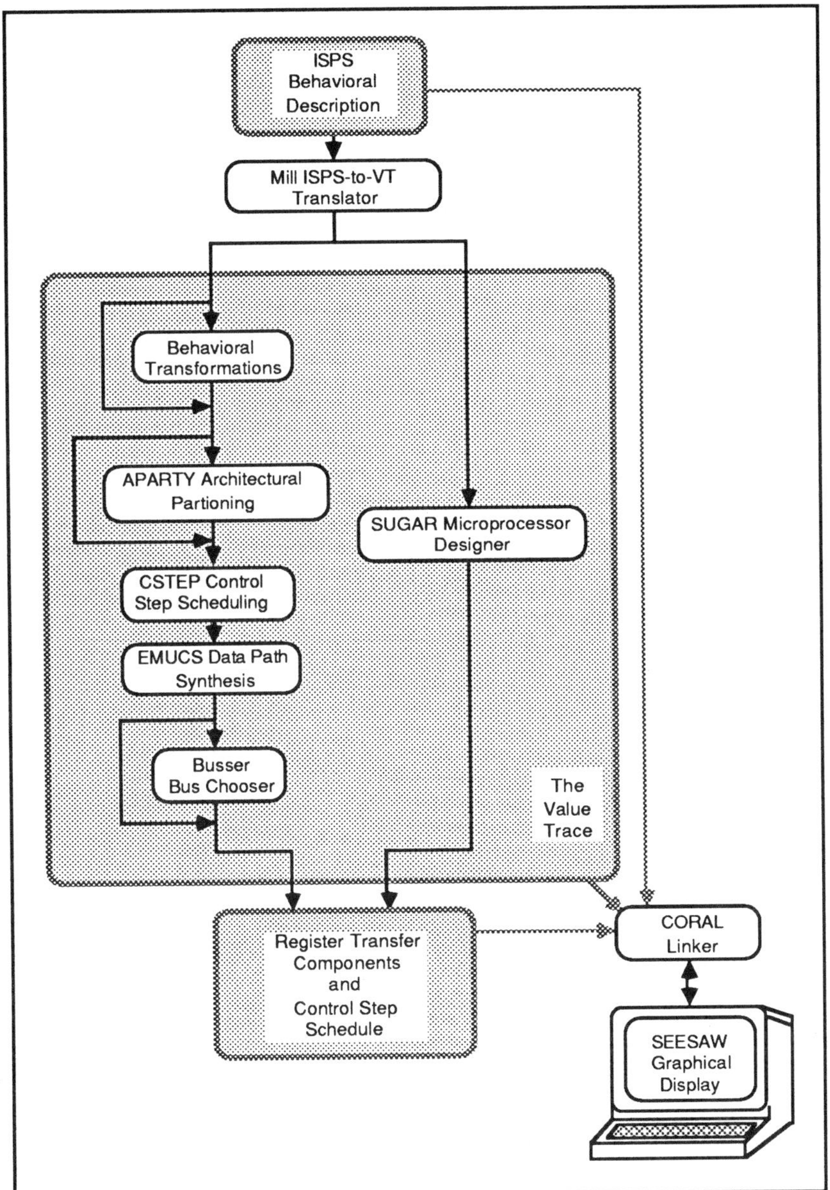

Figure 1-4 . The System Architect's Workbench

- The Architectural Partitioning tool (APARTY) proposes the major structural features of the design from the transformed behavior. The tool suggests how certain behavioral operators should be grouped and implemented by common hardware.

- The Control Step Scheduler (CSTEP) schedules the behavioral operations into control steps, determining the parallelism of the design. It considers the structural partitions suggested by the Architectural Partitioning tool as well as timing and resource constraints specified by the designer.

- The Data Path Allocation tool (EMUCS) determines the number of registers, functional units, and interconnections needed to implement the behavior and assigns appropriate behavioral entities to them. A post-processing phase determines the bussing structure.

- The SUGAR style-specific microprocessor designer includes both behavioral transformation and Register-Transfer Level synthesis functions.

- The CORAL tool maintains the correspondences between the initial behavioral description and the resulting Register-Transfer Level design. This program is shown linked to the SEESAW display for visualization and other purposes.

The steps within Figure 1-4 can be correlated to those within Figure 1-1. The behavioral transformations and architectural partitioning steps in Figure 1-4 correlate to the Algorithmic Level synthesis step of Figure 1-1. CSTEP and EMUCS correspond to the Register-Transfer Level synthesis step in Figure 1-1. The SUGAR microprocessor synthesis program includes aspects of both of these steps.

1.3 CONTRASTING APPROACHES TO SYNTHESIS

The Workbench research can be viewed on several levels. On one level, the tools each represent research into synthesis algorithms and techniques. On another level, contrasting approaches taken by the individual tools allow for interesting comparisons. Finally, the CORAL project suggests a number of spin-off design tools; new approaches to old design tasks that can be taken because of the detailed linkage between abstract behavior and Register-Transfer Level structure.

The research reported here highlights three contrasts in synthesis tool development, namely: how to handle the interactions between the Register-Transfer Level synthesis steps, the difference between style-specific and general synthesis, and the means of representing knowledge in the synthesis algorithms.

1.3.1. Interaction Between Steps in Synthesis

The steps in synthesis are often interdependent and interact intricately. For instance, there is a cyclic dependency between data path allocation and control sequence scheduling. On the one hand, good scheduling requires knowledge of the assignment of behavioral operations to hardware units so that hardware conflicts can be determined. On the other hand, hardware units can be allocated intelligently only if the allocator knows what behavioral operations have been scheduled in the same step. While most synthesis programs solve allocation and scheduling as independent problems [Brewer87, Hitchcock83a, Kowalski84, McFarland86a, Tseng83], some recent synthesis programs [Parker86, Paulin87] interleave allocation and scheduling by solving them as interdependent problems. Behavioral transformations may also interact with scheduling and allocation. For instance, the decision to move some operations out of the branches of a conditional may result in a different control sequence with different hardware requirements. Except for Flamel [Trickey87], most synthesis programs do not consider interactions between behavioral transformations, scheduling and allocation.

When interactions are ignored, decisions made in earlier design steps may lack foresight and prevent good decisions from being made in later design steps. One possible method to reduce harmful interactions is to use an iterative refinement method in which results from a previous iteration are analyzed to either change certain decisions or to change the order in which certain decisions are made. Another possible method is to use heuristics to resolve interactions. The "general synthesis" path in the Workbench uses the Architectural Partitioning results as a top-down guide for the other tools to follow. The SUGAR design tool combines the iterative and heuristic methods with a least-commitment style of decision making called delayed binding. The delayed binding method tries to reduce interactions between design steps by deferring decisions until they can be made in an informed way. Results from the general synthesis path and the SUGAR microprocessor designer will highlight differences in these approaches.

1.3.2 Design Style

Usually, there are several techniques that can be used to implement a given behavior. Each technique exhibits specific structural and organizational features and is characterized by specific design tradeoffs and methods of synthesis. Structures with similar features can be grouped into broad categories called *design styles* [Thomas77, Brewer86].

Synthesis programs and the methods they use can be classified as being style-specific or general-purpose. The style-specific programs view synthesis as a knowledge intensive problem requiring design methods and design tradeoffs specially tailored to individual design styles. CATHEDRAL [De Man86, Rabaey88] is an example of a style-specific program for digital signal processing.

A general-purpose synthesis program uses generally applicable synthesis methods to design a digital system. Instead of directly encoding the knowledge about the design style, information in tables or knowledge data bases along with design analyses is used to guide decision making [Thomas88].

The style-specific programs are fast and produce good results for particular design styles but cannot normally be applied to all aspects of a design since most designs are non-homogeneous and consist of various design styles. For instance, design tradeoffs and methods used to design a microprocessor's control section or interface are different from those used to design its execution unit.

A style-specific method looks for clues in the given problem to determine what features occur in it, retrieves design methods relevant to the features and applies them to solve the problem. Kowalski's interviews with designers [Kowalski84] suggest that expert human designers use a style-specific method. The designers rarely design a structure from scratch; instead, they look for features they have encountered in the past to guide the selection of appropriate structures.

The general synthesis and the SUGAR microprocessor design paths will contrast this aspect of synthesis system organization.

1.3.3. Knowledge Representation

The knowledge representation method is an issue orthogonal to the synthesis method. Most synthesis programs use either purely algorithmic methods [Hitchcock83a, Marwedel86, Parker86, Tseng83] or purely rule-based methods [Kowalski84]. The developers of the former programs believe that it is possible to capture all design knowledge as a few algorithms. The developers of the latter programs believe that all design knowledge can be captured as small chunks of knowledge called rules.

The algorithmic methods produce excellent results for subproblems for which good algorithms exist. However, owing to the complexity of algorithmic and Register-Transfer Level synthesis, many subproblems may not be cast easily as effective algorithms. To make their solution tractable, the algorithmic methods must often make simplifying assumptions. The rule-based techniques are particularly useful in solving subproblems that require the application of many heuristics. Since many rule-based programs use a fixed strategy and make decisions opportunistically in an ad hoc way, they require an extensive and complex knowledge base to compensate for these limitations.

Recently, synthesis programs have been combining the algorithmic and rule-based methods [Brewer87, McFarland86a, Paulin86]. The combined approach has the advantage that the knowledge representation can be tailored to the subproblem to be solved. The SUGAR tool builds on previous research in this area by presenting a combined approach to building synthesis programs. The algorithmic method is used to solve problems for which effective and efficient solutions exist, while the rule-based method is used to solve subproblems whose underlying structure have little regularity and whose design space and complexity are large. The knowledge is represented both as large complex rules and as algorithms. This is in contrast to the Workbench's general synthesis path where all knowledge is represented as algorithms.

1.4. HISTORICAL NOTE

Considerable research in the areas of Algorithmic and Register-Transfer Level synthesis has led to the current state-of-the-art. The EXPL system of Barbacci [Barbacci73] was probably the forerunner of current Register-Transfer synthesis programs. It explored cost/speed tradeoffs using a fixed register-transfer module set by applying serial/parallel transformations to the behavior. The Carnegie Mellon University Design Automation (CMUDA) project [Director81, Thomas83] and The System Architect's Workbench project [Thomas88] are direct descendants of the EXPL system. References to other work will be made in the appropriate chapters.

1.5. OVERVIEW OF THE BOOK

This book describes the coordinated research projects that are part of the System Architect's Workbench. Chapter 2 describes the Workbench's design representations and further describes the Algorithmic and Register-Transfer Level synthesis steps. Chapters 3 through 6 then describe the general synthesis tools in depth. Chapter 7 provides a contrast by presenting the style-specific SUGAR synthesis tool. In Chapter 8 we further contrast the results of these two parallel paths through the Workbench. Finally, Chapter 9 describes a method of linking the original behavioral description with the synthesized Register-Transfer Level design and suggests future design tools that could be developed based on the linked representation.

2. Design Representations and Synthesis

Before describing each of the Workbench design tools in detail, it is important to understand both the conceptual and detailed models of design representation used by the tools. This chapter presents the Workbench's models of design representation, and with this background, defines the synthesis steps implemented by the Workbench tools.

2.1 THE MODEL OF DESIGN REPRESENTATION

The Workbench's model of design representation uses three axes, each representing one of three *domains of description*, to represent a design. Figure 2-1 shows the three domains of description, namely the Behavioral, Structural, and Physical Domains, represented by three axes approaching a common vertex. Along each of these axes, multiple *levels of abstraction*, or *levels of detail*, are represented as *rings* that pass through all three domains.

Other researchers have similar models of design representation; Gajski and Kuhn's *Y-chart* [Gajski83], and Knapp and Parker's *Design Data Structure* (DDS) [Knapp84] are two other examples. The three models are compared in a later section of this chapter.

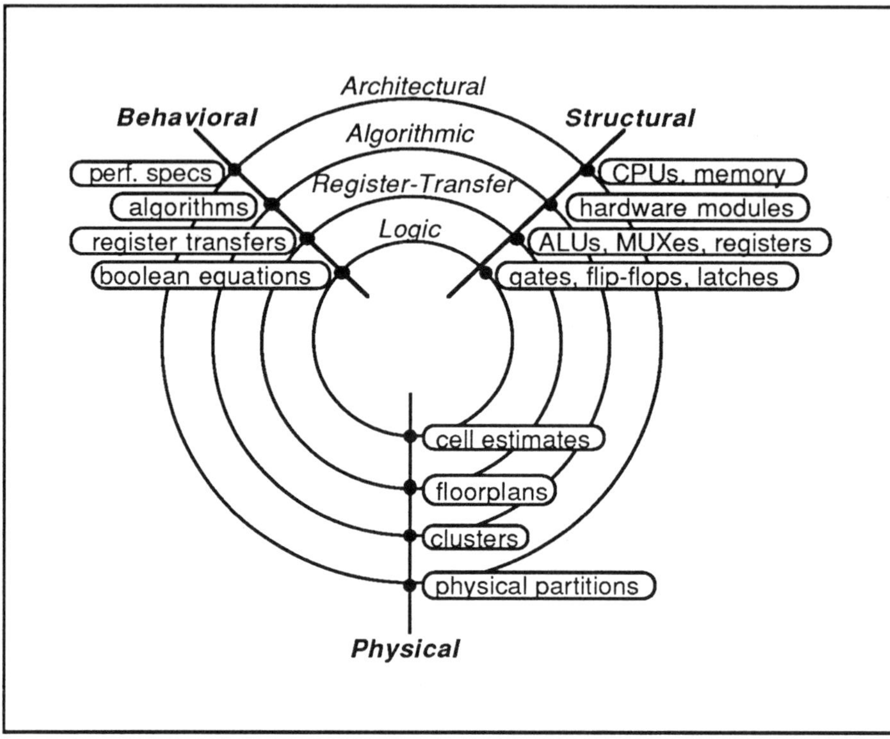

Figure 2-1. The Workbench Model of Design Representation

2.1.1 Domains of Representation

Each of the domains represents its particular portion of the total design description, and must be coordinated with the other domains through inter-domain links. The Behavioral Domain describes the basic functionality of the design, the Structural Domain the abstract logical implementation of the design, and the Physical Domain the physical implementation of the design. Other design decisions, such as those involving timing or topology, are subsumed by the domains and are not represented separately.

The *Behavioral Domain* describes the behavior, or functionality, of the design, either as a complete, program-like description, or as set of behavioral descriptions. For example, at the Algorithmic level, the Behavioral Domain might be represented by a description of the algorithm to be implemented. Variables and the operations on them will be described using standard programming constructs (IF, FOR, etc.) and subroutines. At the lower Register-Transfer level, two types of behaviors will exist. First, behavioral models of each of the Register-Transfer level data path modules (e.g. behavioral models of registers and functional units) are described. Second, a control step schedule for the behavior will indicate the sequence in which the RT level data path modules will be activated. Within any level in the Behavioral Domain there may be a program hierarchy of subroutine calls. However, between Behavioral Domain levels, the relationships are non-hierarchical.

The *Structural Domain* describes the hierarchical logical implementation of the design in an abstract, technology independent fashion. For example, at the Register-Transfer Level the Structural Domain describes the design in terms of an abstract set of functional units, multiplexors, and registers needed to implement the register transfers required by the Behavioral Domain, plus a controller and control ROM.

The *Physical Domain* describes the physical implementation of the design. Like the Structural Domain, the Physical Domain is hierarchically decomposed, although perhaps with a different decomposition structure. At the upper levels of abstraction, this domain contains information on the physical partitioning or floorplanning of the design, and at the lower levels of abstraction, this domain contains a description of real physical components and their geometry (cell detailing). As an example, at the Register-Transfer Level, the Physical Domain describes the floorplanning needed to implement the required abstract functional units, multiplexors and registers. At all levels, speed, power, and area constraints are often part of the Physical Domain.

2.1.2. Characterization of the Domains

Non-isomorphism. The hierarchical decompositions in each of the three domains of description may be non-isomorphic, meaning that there is not a one-to-one mapping between the decomposition structures. For example, Figure 2-2 shows the synthesis of two operators in the Behavioral Domain, an add operator and a subtract operator, with a single Structural Domain components, an arithmetic-logic unit that

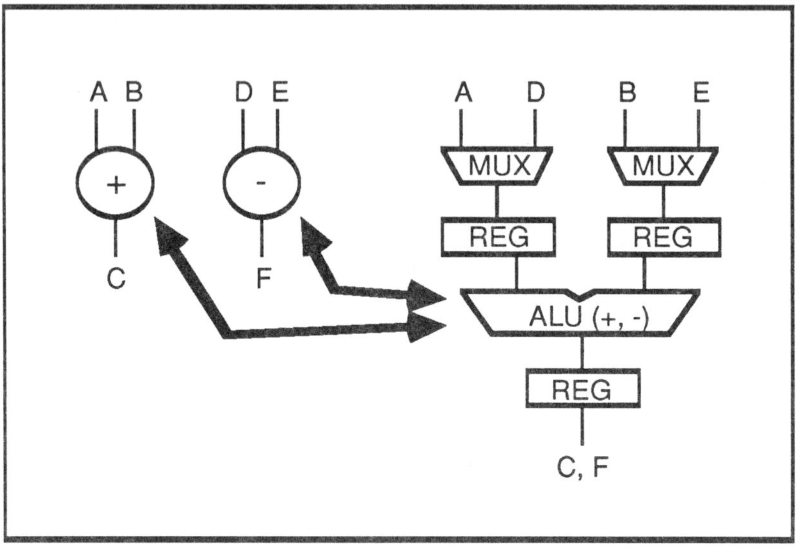

Figure 2-2
Non-isomorphic Behavioral and Structural Domains

can perform both addition and subtraction. Since two Behavioral Domain elements correspond to one Structural Domain element, the two domains are non-isomorphically decomposed. Thus, it is necessary to maintain the design as three separate, but linked, decomposition structures.

Description and Constraint Components. Each of the three domains may also contain both a description component and a constraint component. The *description* is a representation of the design as it exists at that point in the design process. This description may be provided by the user (e.g., as the input description for a synthesis system) or it may be built under program control as an intermediate or final representation of the design and linked to the user-supplied information. The second component of each domain, the *set of constraints*, restrains the design process. These constraints may be represented separately from the description, or they may be combined with the description, possibly as attributes. Timing constraints, area constraints, and power consumption constraints are typical examples.

2.1.3. Multiple Levels of Abstraction

In this model of design representation, there are many *levels of abstraction*, each containing a Behavioral, Structural, and Physical Domain component. Only the levels of abstraction pertinent to the Workbench will be presented in detail.

System Level. At the *System Level*, the Behavioral Domain describes the behavior of a system as a set of performance specifications, or gross operational characteristics. It is concerned with operations that manipulate data and store results, without being concerned with the algorithms describing the details of how this occurs. The Structural Domain components that correspond to these performance specifications are processors, memory units, switches, busses, and device controllers. In the Physical Domain, high-level physical partitioning of the design (e.g. onto cabinets, boards or chips) may be described.

Algorithmic Level. The *Algorithmic Level*, often called the Behavioral Level, describes the design at a level syntactically similar to programming languages. Complex algorithmic expressions, data structures, procedures, processes, and scoped variables are all used to describe the behavior of the design. The behavior of the system is modeled as a set of *concurrent* programs. Each of the programs describes an independent thread of control, or *process*, which will most likely be implemented with a separate controller and data path. For example, the Behavioral Domain at this level might describe the instruction decoding and execution of a particular instruction set as a single process, and it might also describe a separately controlled prefetch unit that runs concurrently with the instruction execution.

In the Structural Domain, hardware subsystems (e.g., separate data paths and controllers) represent the processes of the Behavioral Domain. These hardware subsystems are thought of as operating concurrently with each other, although their internal operation may include both sequential behavioral and statement-level parallelism.

In the Physical Domain, clustering of operators into physical subsystems might be described. Clustering, or partitioning, refers to grouping functionally similar operations together as determined by some measure of *proximity*. Besides this description component, constraints on power consumption, area, etc. may also be specified.

Register-Transfer Level. The *Register-Transfer Level* describes the design at a level much closer to the underlying hardware than the previous levels. Processes are each implemented as controller - data path pairs. For each of the concurrent processes, this level describes the register-transfer behavior in each state of each controller in terms of the arithmetic and logic operations evoked, the registers loaded, and the next state. When several registers are loaded in the same state, the register transfers are said to be *parallel*.

In an abstract fashion, the behavior at this level is implemented in the Structural Domain using ALUs, adders, multiplexors, registers, and memories.

Depending on the target technology, it may be possible to implement this structural description directly in hardware. If this is not the case, or more detail is desired, the Physical Domain at this level might be concerned with floorplanning the layout of the design by specifying the geometrical arrangement of the Structural Domain components. Finally, as with the upper levels of abstraction, power and area constraints may also be part of this domain.

Logic Level. The *Logic Level* is a more detailed version of the Register-Transfer Level. In the Behavioral Domain, it describes the design as a switching circuit, expressing the behavior in terms of boolean equations and finite automata. In the Structural Domain, this behavior is implemented by gates, flip-flops, PLA's controllers, and registers. The Physical Domain represents the partitioning of the structural elements onto finer grained physical partitions.

There are other lower levels that can be represented in our model. However, they will not be presented since the Workbench does not directly deal with them.

2.1.5. Other Models

Other researchers have models of design representation that are similar to the one described here in that they represent multiple levels of abstraction and they represent the design at many phases of the design process. This section describes two of the more prominent of those models.

Gajski and Kuhn. In their introduction to an issue of *Computer Magazine*[Gajski83], Gajski and Kuhn used a tripartite representation of design, called a *Y-chart*, to illustrate the design tasks in various systems. The *Functional Representation*, *Structural Representation*, and *Geometrical Representation* in their model correspond roughly to the Workbench model's Behavioral Domain, Structural Domain, and Physical Domain, respectively, although the Workbench model's Physical Domain also includes non-geometrical information (e.g., actual area and power consumption).

USC's ADAM System. At the University of Southern California, the ADAM System [Granacki85] uses a data structure called the *Design Data Structure* (DDS) [Knapp84] to represent the design. This data structure was developed to provide a common, uniform representation for a set of design automation tools (the ADAM System), and was developed to provide a set of mutually orthogonal *subspaces*. The four subspaces used in the DDS are the data flow behavior subspace, the structural subspace, the physical subspace, and the timing and control subspace. Taken together, the *data flow behavior subspace* and the *timing and control subspace* correspond roughly to the Workbench model's Behavioral Domain. The data flow behavior subspace represents functional definitions and data dependencies between values as a data flow graph. The timing and control subspace, also called the sequencing subspace, represents the ordering of events, the conditional sequencing of events, and the time between events. The third subspace, the *structural subspace*, represents the electrical topology of the design, and is similar to the Workbench model's Structural Domain. Likewise, the *physical subspace* is similar to the Workbench model's Physical Domain, encompassing both geometrical and purely physical information.

2.2 . BEHAVIORAL REPRESENTATIONS AT THE ALGORITHMIC LEVEL

The model described in the previous section provides a philosophical backdrop upon which more detailed representations and synthesis steps may be described. The rest of this chapter begins to present these details. First, the Algorithmic level behavioral representations used as input to the Workbench tools are presented. Second, the Register-Transfer representations generated by the Workbench are presented. With this more-detailed background, the synthesis steps necessary to generate the Register-Transfer level from the Algorithmic level representations are defined.

The ISPS language provides the starting Algorithmic level Behavioral Domain specification for the Workbench tools. A control and data-flow graph representation, called the *Value Trace* (VT) [Snow78] is generated from the ISPS. The VT is a better representation for synthesis because it allows the removal of language artifacts from the program-like behavior. Thus it provides a less stylized representation of the behavior. This section summarizes the characteristics of ISPS and describes the VT in detail.

2.2.1. ISPS

The Instruction Set Processor Specification (ISPS) language [Barbacci80, Barbacci81], based on the earlier ISP notation [Bell71], is a computer hardware description language for describing the behavior of System level components. ISPS has been used as a vehicle for a variety of tasks [Parker79] including description and documentation of computer hardware, as well as simulation and architectural verification. In the Workbench, it is employed for describing the behavior of a piece of hardware to be designed.

Language Description. The basic memory element in ISPS is the *carrier*. A carrier has a name, a word structure, and a bit structure, and represents either architectural or temporary registers and arrays. In the ISPS description in Figure 2-3, the carriers include **i**, **j**, and **vi**. Carriers **i** and **j** have bit structures, both <3:0>, declaring them to be 4 bits wide and **vi** is declared to be 16 bits wide. **Anew** has a word structure ([0:255]), declaring it as an array of words numbered from 0 to 255, and a bit structure (<15:0>), declaring it to be 16 bits wide, numbered from 15 to 0.

Values are transferred from one carrier to another through statements describing transfer operations and data operations, much the same as in any other procedural language. In the example, the statement:

```
x[j] = 0;
```

indicates a memory write into carrier **x**, to the location specified by carrier **j**. The value written is the constant 0.

Statements are separated in one of two ways. First, the NEXT construct specifies that all the statements before it must execute before any of the statements after it may start. Second, the semicolon (";") construct specifies that execution of the ISPS statements may proceed in an arbitrary order as long as the lexical data precedences are maintained. Note that this does not conform to the original ISPS language and simulator definition which specified semicolon to mean order independence regardless of the lexical data precedences. All examples in this book follow the revised definition.

ISPS also supports procedures and procedure calls; the ISPS Reference Manual refers to a procedure as an *entity* with a *behavioral expression* for a body, and to a procedure call as the *activation* of an entity. As in most procedural languages, procedures can be passed a set of parameters (values), and a single value can be explicitly returned by the procedure. Furthermore, values may be transferred to the procedure through global carriers and modified, causing side effects.

ISPS provides several looping constructs. The REPEAT operation, shown in the example, specifies that the contents of the following block of statements (delimited by BEGIN and END operations) are to be executed repeatedly. Another operation, the RESTART operation, can be used at the end of a labeled block of statements, also causing a block of statements to be execute repeatedly. Once started, a labeled block of statements can be left via LEAVE, TERMINATE, RESTART, or RESUME operations; only the LEAVE statement is shown in this example.

ISPS provides two mechanisms for conditional execution. IF operations evaluate an expression, and execute an associated statement or block of statements if that expression evaluates true (non-zero). DECODE operations, similar to the SWITCH or CASE statements in other languages, evaluate an expression, and execute one of a set of alternate statements or statement blocks, depending on the value of the expression.

ISPS also supports a *qualifier* mechanism to supply arbitrary additional information to applications programs. Qualifiers consist of an identifier and a list of values surrounded by "{}" brackets, and can be attached to any operation, carrier, or entity.

```
kalman.isp := BEGIN
    ** carriers **
    i<3:0>,
    j<3:0>,
    vi<15:0>,
    tempi<15:0>,
    Anew[0:255]<15:0>,
    k[0:255]<15:0>,
    g[0:127]<15:0>,
    x[0:15]<15:0>,
    y[0:12]<15:0>,
    v[0:3]<15:0>,

    ** statements **
filt{main} := BEGIN
    j = "F;
    s1 :=
      REPEAT BEGIN
            x[j] = 0;
            IF (j = j - 1)<4> => LEAVE s1
      END;
    i = "F;
    s2 :=
      REPEAT BEGIN
            j = "F;
            tempi = 0;
            s21 :=
              REPEAT BEGIN
                    tempi = tempi + (Anew[i@j] * x[j])<15:0>;
                    IF j LEQ "C =>
                            tempi = tempi + (k[i@j]*y[j])<15:0>;
                    IF (j = j - 1)<4> => LEAVE s21
              END;
            x[i] = x[i] + ("FFFF * tempi)<15:0>>
            IF (i = i - 1)<4> => LEAVE s2
      END;
    i = "7;
    s4 :=
      REPEAT BEGIN
            vi = 0;
            j = "F;
            s41 :=
              REPEAT BEGIN
                    vi = vi + (g[i@j] * x[j])<15:0>;
                    IF (j = j - 1)<4> => LEAVE s41
              END;
            v[i] = (vi * y[i])<15:0>;
            i = i - 1;
            IF i LEQ "3 => LEAVE s4
      END
    END
END
```

Figure 2-3. An ISPS Example

The system being described in Figure 2-3 is a Kalman filter, a real-time controller consisting of several loops implementing matrix multiplication and other operations. The input to the system is the **y** array, and the output is the **v** array. First to be described are the carriers that will take part in the computation. Second, the procedures that specify the actions on the variables are listed. In this case, there is only one procedure (**filt**) consisting of several nested, labeled blocks and repeat loops. The "**{main}**" qualifier indicates that **filt** is the starting point for execution. A quote mark (") indicates a hexadecimal constant, labeled blocks are named with a colon-equal (:=), and the "@" is the binary concatenation operator.

Language Usage. Following the conventions described in our model of representation, the ISPS behavioral domain description is maintained as a pure behavior, represented separately from the structure or physical information. That is, the use of a "+", "-", or "*" operator in ISPS only specifies that the operation must take place on the named values. It does not specify the actual functional units to be used for each of the operations, the memory units in which the values are to be stored, nor does it specify the order in which the operations will take place. Thus, the Workbench has the freedom to design any hardware and control sequence implementation with the proviso that the final implementation's behavior match that specified by the lexical data-flow ordering and the required (NEXT) ordering of the original behavior.

Prespecification of structure in the Workbench is partially supported by allowing the designer to enter some of the Register-Transfer level structure description and then use interdomain links to specify what part of the behavior it will implement. (Currently, only the EMUCS data path synthesis tool provides this capability.) For instance, two separate functional units could be specified to implement two separate ISPS operators; say, one could do the subtract of value **j** in the **s1** loop of Figure 2-3 and the other could do the subtract of value **j** in the **s2** loop. This specification does not change the behavior by cluttering it up with implementation specific detail. But it does give the designer the capability of suggesting part of the implementation and letting the Workbench fill in the rest.

Chapters 3 and 9 describe the interdomain links between the behavior and structure in more detail and Chapter 9 also illustrates how they may be used to interrogate the automatically synthesized design.

2.2.2. The Value Trace (VT)

The Value Trace [Snow78] is an internal control and data-flow graph representation of ISPS. This graph represents behavior in terms of *operators* that correspond to ISPS operators and the *values* that pass between them. Operators are represented by *nodes* in the graph. They perform a function on their inputs and produce one or more outputs. Operator inputs and outputs are connected to other operator inputs and outputs by directed *edges* that represent values. Each value represents an individual value of an ISPS variable or intermediate expression. Since variables may be assigned several values in an ISPS description, there may be several values for each ISPS variable.

The value edges form a partial ordering of operators - edges describe the precedence relationships between operators without specifying a full ordering. In addition, while value edges represent the values of variables, they do not constrain values representing a variable to be stored in the same place in a design. These features remove some of the language artifacts of a textual procedural language such as ISPS and make the VT a more flexible representation for exploring design tradeoffs.

Figure 2-4 shows an example VT graph produced from the ISPS statements:

```
A = A + B;
B = A - C;
```

This graph has two operator nodes that represent the addition and subtraction operators. Value edges describe the flow of values that represent the different values of **A**, **B**, and **C**.

As illustrated by the figure, the following notation is used to identify the components of the VT. Operator nodes are described by the set:

$$X = \{ x_a \}$$

where each operator x_a is indexed by subscript a. When describing operator inputs and outputs it is necessary to distinguish them from each other. This is accomplished using the sets:

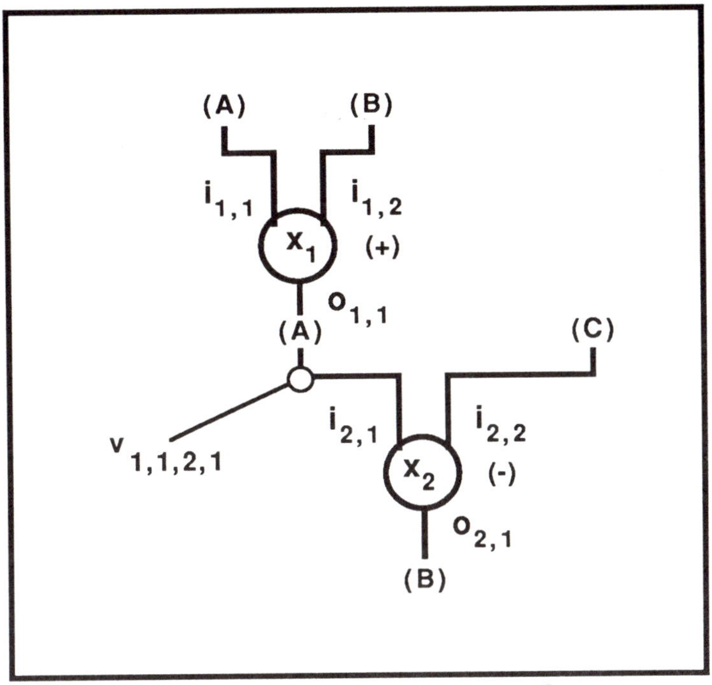

Figure 2-4. Simple VT Example

$$I = \{ i_{a,b} \}$$

$$O = \{ o_{a,c} \}$$

to describe the sets of all operator inputs and outputs, respectively. Each operator input is indexed by the operator index a and a second index b that numerically identifies the inputs to the operator. Similarly, each operator output is specified by the operator index a and a second index c that numerically identifies the operator's output.

Value edges are described by the set:

$$V = \{ v_{a1,c1,a2,b2} \}$$

where each value edge describes a connection between an operator output $o_{a1,c1}$ and an operator input $i_{a2,b2}$

This representation is sufficient to describe ISPS specifications containing a single straight-line sequence of statements with no changes in control flow. Additional features are necessary to describe and support control constructs such as conditional execution, loops, and procedures.

The VT describes control flow by grouping operators into *basic blocks*. Each basic block represents a group of operators translated from a straight line sequence of ISPS statements (a block of statements containing no branches) and a single *control operator* that determines flow of control after all of the operators in the block are executed. Operators within a basic block may execute in any order that does not violate data dependencies. Flow of control between basic blocks is represented by a set of *precedence edges* that specify which blocks may be executed following the completion of a given block.

We use the following notation for basic blocks and precedence edges:

$$B = \{B_i\}$$

describes the set of basic blocks and

$$PC = \{pc_{i1,i2}\}$$

describes the set of precedence edges, where each edge $pc_{i1,i2}$ describes possible flow of control between two basic blocks B_{i1} and B_{i2}.

Control operators are a special set of operators that specify which precedence edge is taken when leaving a basic block. In addition to specifying control flow, control operators also specify the data flow characteristics of the control action: which values are used, and which new values are produced as a consequence of the control action. Specific control operators support conditional control flow, loops, and procedures.

The SELECT control operator supports conditional control flow by providing a multiway branch. A SELECT operator uses an input value to transfer control to one of two or more branches, where each branch is composed of a set of basic blocks. Control is transferred back to the SELECT operator after the basic blocks in the branch have executed.

SELECT operators produce output values that explicitly represent the effect of conditionals on data flow. Each output represents the value of a variable that is changed in the branches. Explicit SELECT outputs maintain consistency between variables that are conditionally changed in branches so that either a new value is produced or an old value is maintained.

Figure 2-5 illustrates the control flow and data flow aspects of a simple SELECT operator that is translated from the ISPS statement:

```
IF (A) => I = I + 1
```

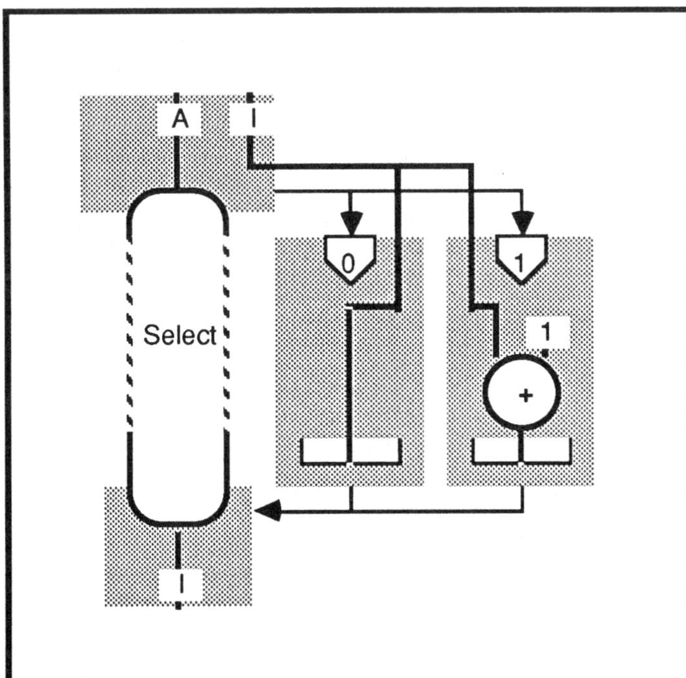

Figure 2-5. VT Conditional Control Flow

This statement increments variable **I** if expression **A** evaluates to a nonzero value. Basic blocks are shown by the shaded areas, and control flow edges between basic blocks are represented as thin lines. The SELECT operator is split into upper and lower parts to illustrate the two transfers of control involved in the SELECT. The top part represents transfer of control to one of the branches, while the lower part represents transfer of control to the end of the SELECT operator after the branch completes execution. This SELECT operator has two branches, where each branch consists of a single basic block. One branch contains an addition operator while the other branch contains no operators. The data flow aspect of the SELECT operator is illustrated by the branches and lower part of the SELECT operator. Depending on which value is changed, the variable **I** is either incremented or left unchanged. To maintain consistency, the old value and new value of **I** both feed the same SELECT output. This output value is transferred from the value in the branch when the branch completes execution.

When all of the branches of a SELECT operator contain no operators, no change in flow of control is needed at all. In this case, only the data flow part of the operator matters and the operator can be implemented using a multiplexor. This special case SELECT operator is known as a *data SELECT*.

Procedures are supported in the VT by grouping a set of basic blocks into a subgraph called a *vtbody*. A vtbody is a graph that is translated from an ISPS procedure or labelled block. Values from surrounding code that are used in the ISPS procedure become explicit data-flow inputs to the graph. Similarly, values that are produced in the procedure and referenced in the surrounding code become explicit data-flow outputs of the vtbody graph.

Control is transferred to a vtbody through either a CALL or ENTER control operator depending on whether the vtbody represents a procedure that is called or a labelled block, such as **s1** in Figure 2-3, respectively. Both operators have inputs and outputs that correspond to the vtbody graph's inputs and outputs. When control is transferred to a procedure or labelled block, the inputs are transferred to the vtbody inputs. Procedure and labelled block exits are supported with the LEAVE operator. A LEAVE operator transfers control back to the calling environment. The transfer of vtbody outputs to CALL or ENTER operator outputs is explicitly represented by inputs to the LEAVE operator - each LEAVE operator has a number of inputs that matches the number of vtbody outputs.

Loops are represented as a special case of vtbodies. A loop is represented as a vtbody that is activated and terminated as above. The use of the RESTART operator allows iteration to occur. A RESTART operator has the same set of inputs as a CALL operator and passes control back to the beginning of the loop with a new set of values.

The notation used for vtbodies is described as follows. The set:

$$VTB = \{ vtb_m \}$$

describes the set of vtbodies; each vtbody is described as a set of vtbody inputs and a set of basic blocks of operators.

The source of inputs to CALLed vtbodies is only known during the execution of a specification. For this reason, vtbody inputs are described as edges that originate from a dummy operator x_0 :

$$O_0 = \{ o_{0,c} \}$$

These edges are in turn described as the set:

$$V_0 = \{ v_{0,c,a,b} \}$$

Figure 2-6a shows an example ISPS description and Figure 2-6b shows the corresponding VT diagram, including a loop vtbody and the ENTER operator that activates it. The ENTER operator takes as inputs values for **I** and **DATA** which are passed to the vtbody inputs for the same value. Each time the loop restarts, the current values of **I** and **DATA** are passed back to the vtbody for use in the next iteration. When the loop terminates, the final values of **I** and **DATA** are passed back to the outputs of the ENTER operator.

```
example := BEGIN
** Variables **
data<7:0>,
i <3:0>
** Statements **
exmain := BEGIN
    i = 0;
    exloop := BEGIN
       data = data SR0 data ;
       i = i + 1;
       IF ( i LEQ 8 ) RESTART exloop;
       LEAVE exloop
       END
    END
END
```

Figure 2-6a. An Example ISPS Description

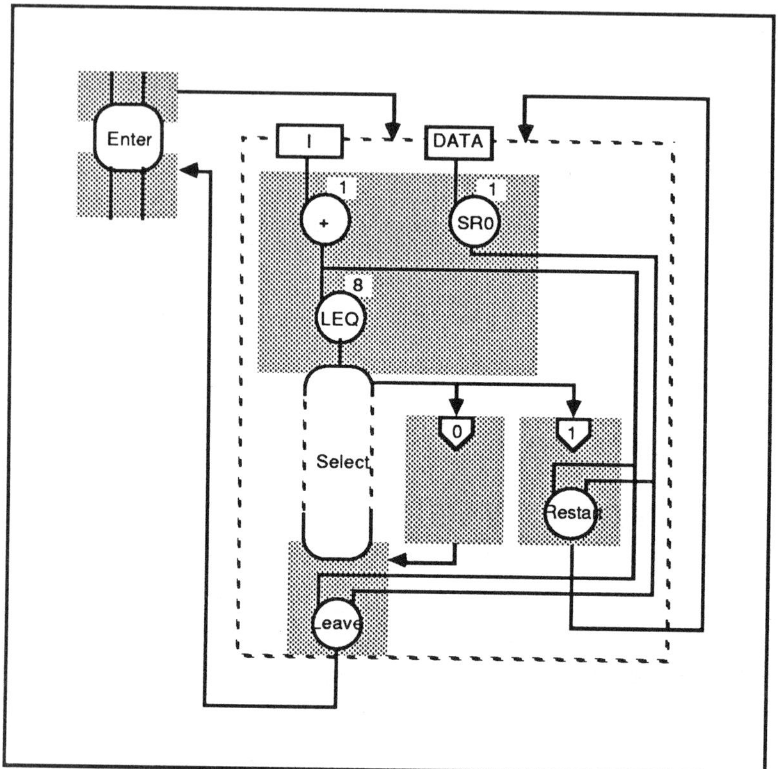

Figure 2-6b. A VT Containing a Loop

2.3. BEHAVIORAL AND STRUCTURAL REPRESENTATIONS AT THE REGISTER-TRANSFER LEVEL

Register-transfer synthesis translates a behavioral specification into a Register-Transfer level hardware implementation. This implementation is represented by a set of *modules* and *interconnections*. This section describes the representation of hardware used by behavioral synthesis tools.

The implementation of a nonconcurrent digital system consists of a single controller and a data path. Following synthesis the controller is represented as an abstract state machine, with a set of control steps that sequence the operation of the data path. The data path is represented as a

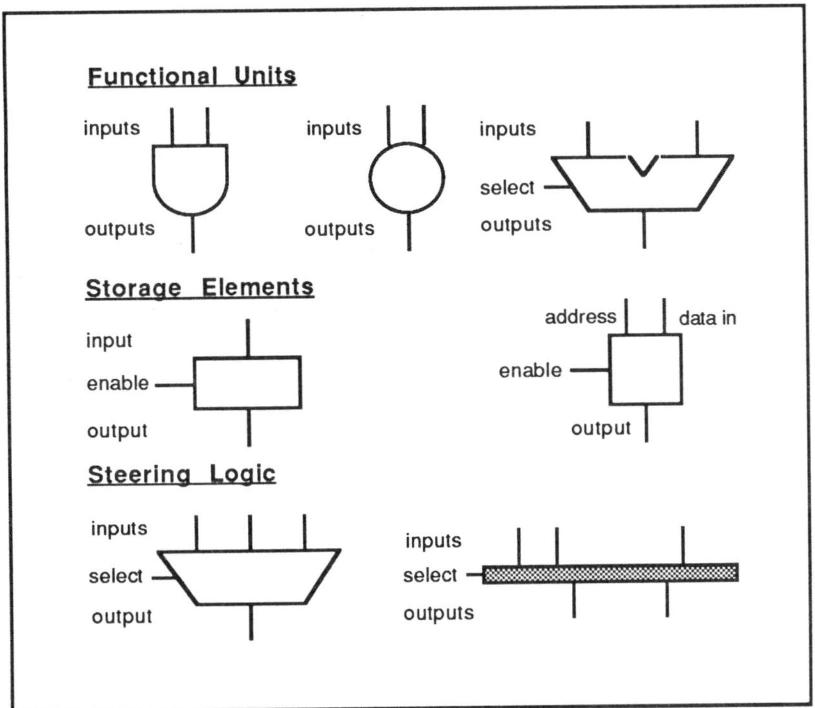

Figure 2-7. Data Path Primitives

set of primitive modules of three different types: functional units, storage elements, and interconnection logic. Figure 2-7 depicts these module types.

Functional units are combinational logic elements that implement operators in the data path. Some functional units perform single functions while others perform multiple functions. Multiple function units use a *select* input to determine which function to apply. The select input is usually connected to the controller, which allows the controller to determine which operators are performed by the functional unit in each control step.

Storage elements include registers and memories. Registers store the value on the *input* port when an *enable* input is asserted and maintain this value on an *output* port until a new value is stored. The enable input of a register is usually connected to the controller, which allows the controller to determine the control steps in which it is loaded.

Memories store arrays of values. Values are stored in the array through a write operation which stores the value on the *data input* in the memory location referenced by the value on the *address input*. Values are read from the array through a read operation that provides a stored value referenced by the *address input* on the *data output*. The *enable* input of the memory invokes the read and write operations.

Interconnection (or *steering*) *logic* includes multiplexors and busses. Multiplexors allow one of several inputs to be selected and passed to a single output. Busses allow one of several inputs to be selected and passed to multiple outputs. A *select* input determines which input and output are selected in a multiplexor or bus. This input is connected to the controller, which allows the controller to specify how data is steered through the data path in each control step.

When connected together and combined with a controller, these modules form a Register-Transfer level design that implements a behavior. The controller is represented as a state machine that sequences the data path through abstract clock cycles called *control steps*. During each control step, information available at storage element outputs passes via links and steering logic to functional units that perform combinational operations. The outputs of the functional units pass through other links and steering logic to register inputs or controller inputs. Register inputs store this information for use in later control steps, while controller inputs determine the sequencing of operations in later control steps. Each control step contains a set of *micro-operations*. A micro-operation is an abstract representation of the select and enable inputs required to either implement a data operation in a data path module or a control operation in the controller.

Figure 2-8 shows an example of a controller and data path and a set of control steps with micro-operations. The data path consists of a single functional unit, three registers, and a single multiplexor that selects the right input of the functional unit. Micro-operations in the first control step add the contents of registers R1 and R2, storing the result in R1. Micro-operations in the second control step subtract R3 from R1, again storing the result in R1. In other figures throughout the book, control inputs such as a functional unit's select and a register's enable are not shown but assumed to be present.

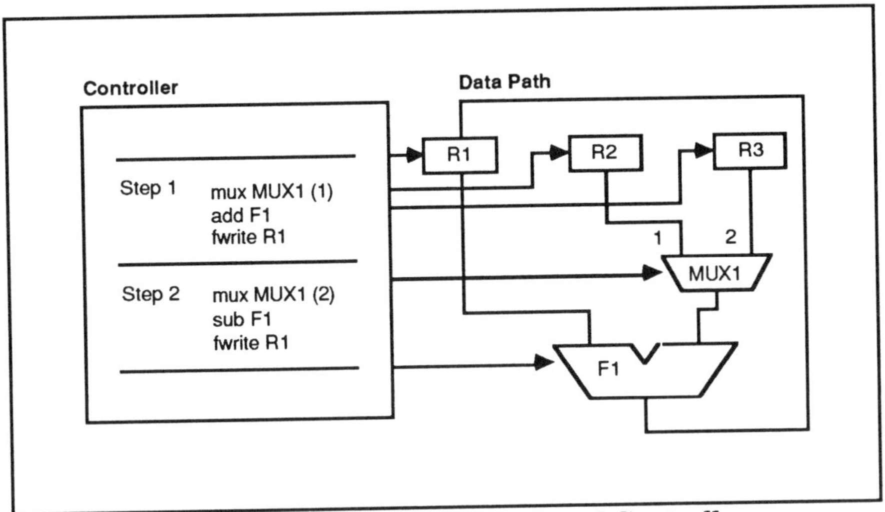

Figure 2-8. Example Data Path and Controller

The following notation will be used to refer to modules, interconnections, control steps, and micro-operations. A design consists of a set of modules:

$$M = \{ m_h \}$$

The set of modules is sometimes broken down by type into four sets of modules:

$$MF = \{ mf_h \}$$

$$MS = \{ ms_h \}$$

$$MW = \{ mw_h \}$$

$$MC = \{ mc_h \}$$

This includes the set of functional units MF, the set of storage elements MS, the set of interconnection steering logic elements MW, and the set containing the controller MC.

Modules are connected by a set of ports:

$P = \{ p_{h,j} \}$

where h is the module subscript and j is the port subscript.

Module ports are connected together by a set of links:

$L = \{ l_{h1,j1,h2,j2} \}$

Control steps are represented by the set:

$CS = \{ cs_{i,l} \}$

where index i identifies the basic block the control step belongs to and index l identifies the control step within the block. These numeric indices only serve to identify the basic blocks and their control steps; next state sequencing is represented in the detailed schedule.

Control step micro-operations are identified by the set:

$MO = \{ mo_{i,l,m} \}$

where indices i and l identify the basic block and control step the micro-operation belongs to and index m identifies the individual micro-operation type.

2.4. MODELING ALGORITHMIC AND RT LEVEL SYNTHESIS

Now that the basic representations for the Workbench synthesis tools have been defined, this section defines the synthesis steps necessary to produce a controller and data path represented at the Register-Transfer level from a behavioral specification at the Algorithmic Level.

Algorithmic level synthesis consists of transforming an algorithm and splitting it among the Structural and Physical domain partitions of a system. Register-Transfer level synthesis consists of two major interrelated phases: control step scheduling, and data path allocation. Conceptually, control step scheduling synthesizes a Behavioral Domain description at the Register-Transfer level from the Algorithmic level behavior, and allocation generates Register-Transfer level structure from the behavior. These steps are illustrated in Figure 2-9.

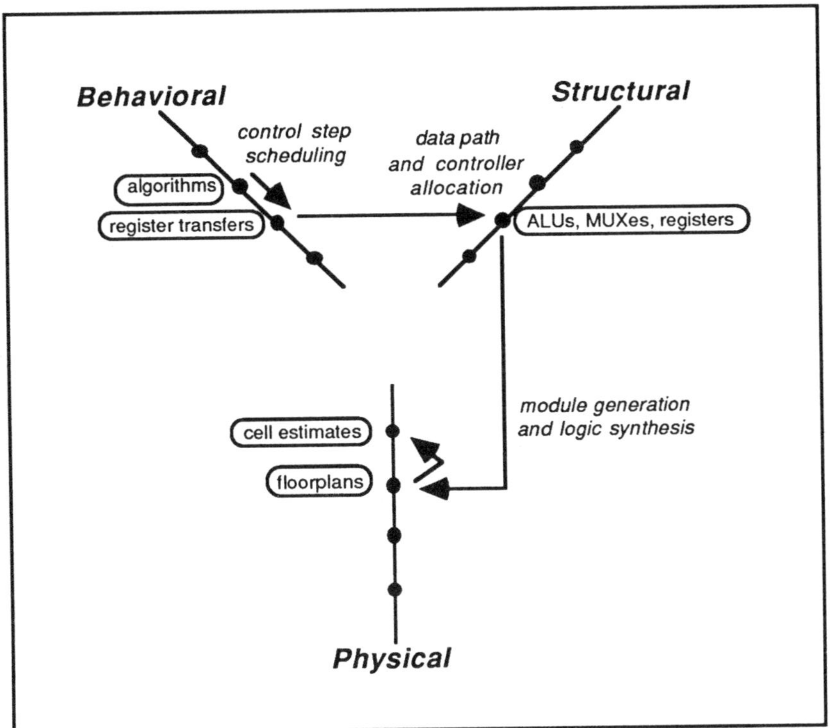

Figure 2-9. Modeling Algorithmic and RT Level Synthesis

2.4.1. Algorithmic Level Synthesis

Algorithmic level synthesis is the process of transforming and/or changing the algorithm to be implemented. As illustrated earlier, this may involve adding or deleting functionality in a pre-existing algorithm, or splitting an algorithm among several Physical domain partitions.

The process of adding or deleting functionality is not merely "writing an algorithm to be implemented." Rather it involves a close interaction and iteration with the lower level synthesis processes. At the Algorithmic level, a designer must determine whether the inclusion of certain functionality is worth the physical costs involved in its implementation. A negative answer involves removing a piece of the algorithm. The resulting system will have less functionality but will be implementable in the available technology.

Further, the designer must determine whether functionality must be added. In the example of Figure 1-3, a queue was needed to buffer the arrival rate of information to an algorithm. The need for the queue might be determined by high level simulation of the system or through an iterative design cycle. In either case, the algorithm to be implemented must be changed. These algorithmic changes are shown in Figure 2-9 as a circular arc starting and ending in the Behavioral Domain at the Algorithmic level.

The process of partitioning a behavior potentially impacts all three domains. In the Behavioral domain it may involve changing the level of concurrency in the algorithm or partitioning the behavior into pipeline stages. The partitioning also determines the top of the hierarchy in both the Physical and Structural domains and assigns pieces of the behavior to the major elements in these domains. This partitioning is illustrated by the arcs from the Behavioral Domain at the Algorithmic level to the Structural and Physical Domains at the same level. The rest of the lower level synthesis processes must then fill in the physical and structural hierarchies. When the physical constraints cannot be met by this partitioning, a new partitioning must be proposed based on the new detailed results of lower level synthesis.

2.4.2. Register-Transfer Level Synthesis

Register-Transfer Level Synthesis includes two phases, control step scheduling and data path allocation and binding.

Control Step Scheduling. *Control step scheduling* is the process of creating control steps and assigning VT operations to control steps, effectively creating a new behavioral representation that specifies the register transfers at the Register-Transfer level. Control steps will eventually be implemented as states in a finite state machine or as microwords in a microcoded controller. Scheduling is illustrated in Figure 2-9 by the arc in the Behavioral Domain from the Algorithmic level to the Register-Transfer level.

One or more control steps are created for each basic block B_i. Each operation x_a in B_i is assigned to either a single control step $cs_{i,l}$ or a contiguous range of two or more control steps $[\ cs_{i,l1}, cs_{i,l2}\ ]$. Scheduling affects the outcome of synthesis in several ways. First, it fixes the order and duration of each operator's execution as it is implemented in the design. Second, it fixes the number of operators that execute in parallel in the design and sets a lower bound on the number of functional units needed to implement the design. Third, it specifies the *lifetimes* associated with operation outputs and value edges. Like the lifetime of a variable, the lifetime of an operation output is the range of control steps over which it must be stored. These lifetimes in turn set a lower bound on the number of registers needed to implement the design.

For example, Figure 2-10 shows a sample VT that has been scheduled into control steps. Two add operations are scheduled to execute in parallel in the first control step, so a minimum of two functional units will be required in the design. Value edges that pass between control steps must be stored in registers. In this example, a minimum of four registers will be required in the design.

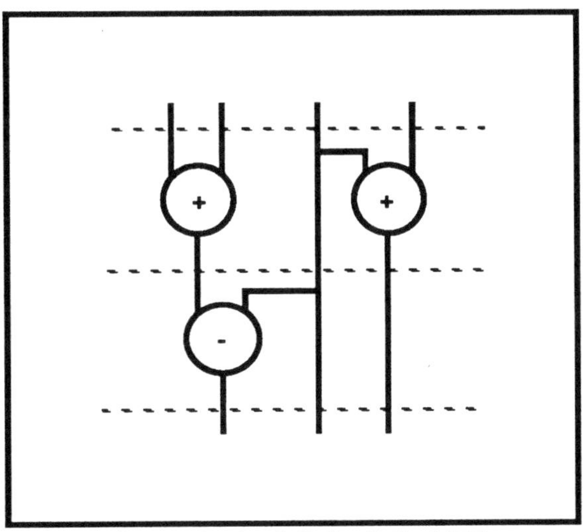

Figure 2-10. A Scheduled VT

Data Path Allocation and Binding. *Allocation* is the process of creating a set of modules M and a set of links L that implement a design as illustrated in Figure 2-9 by the arc between the Algorithmic level of the Behavioral Domain and the Register-Transfer level of the Structural Domain. The modules in a design are usually chosen from a set of predefined abstract modules representing the different module types: *functional units, storage elements, steering logic,* and *controllers*. This available set of modules is technology dependent. Cost and timing information about this set of modules is usually stored in a library database for each technology that is used to guide synthesis decisions.

Binding is the process of assigning the operations and values in the VT graph to allocated modules and interconnections. Binding can be broken down into functional unit bindings, storage element bindings, and interconnection bindings.

A *functional unit binding* assigns an operation x_a to an allocated functional unit mf_h that can perform the operation. Many operations may be bound to the same functional unit if they are not scheduled during the same control step. A functional unit binding includes the creation of a micro-operation that selects the proper function in the proper functional unit.

A *storage element binding* assigns an output value $o_{a,c}$ to a storage element ms_h that will store it for the lifetime of the value. This binding is necessary only if $o_{a,c}$ is used in a different control step than the one in which it is produced. Multiple values may be bound to the same storage element providing their *lifetimes* (the ranges of control steps in which they require storage) do not overlap. A storage element binding includes the creation of a micro-operation that loads the value into the storage element at the end of the control step in which it is produced.

An *interconnection binding* assigns a value edge $v_{a1,c1,a2,b2}$ between operations x_{a1} and x_{a2} to a set of interconnections and steering logic modules that will implement a connection between the appropriate hardware modules in the design. This connection consists of links and steering logic elements that are not already in use during the control steps that the value edge is active.

There are two distinct cases of interconnection binding as illustrated in Figure 2-11. For each case, assume that x_{a1} is bound to the top functional unit and x_{a2} is bound to the bottom functional unit. In the first case (on the left side of the figure), the output of functional unit x_{a1} is alive only in the control step that it is produced and does not need to be stored. In the second case (on the right side of the figure), the output of functional unit x_{a2} is alive for multiple control steps and does need to be stored.

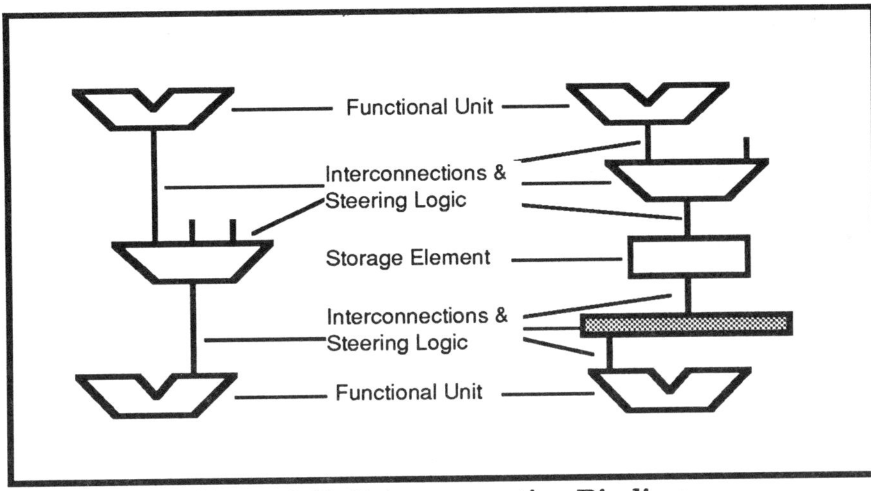

Figure 2-11. Interconnection Bindings

When the value does not need to be stored, interconnections and steering logic in the implementation must form a path from the *output* of the functional unit bound to x_{a1} to the *input* of the functional unit bound to x_{a2}. This path can be a direct link between the two functional units or include steering logic elements. If steering logic is included in this path, then micro-operations that select the proper steering logic inputs and outputs must be added to the control steps in which the transfer of data takes place. The left side of Figure 2-11 shows an example of this case in which two links and a single multiplexor form the necessary path between functional units.

Values that need to be stored are called *regvals*. For regvals, two paths of connection must be established: a path from the *output* of the functional unit bound to x_{a1} to the *input* of the storage element bound to $o_{a1,c1}$, and a path from the *output* of the same storage element to the *input* of the functional unit bound to x_{a2}. Each of these paths can be a direct link or can include steering logic. If steering logic is included, then micro-operations that select the proper steering logic inputs and outputs must be added to the control steps in which data is steered to the storage element input and the control steps in which data is steered from the storage element outputs to the functional unit input. The right side of Figure 2-11 shows an example of this case in which two links and a multiplexor connect the first functional unit to the storage element input, and two links and a bus connect the storage element output to the input of the second functional unit.

2.6 SUMMARY

Now that the conceptual and detailed models of design representation used by the Workbench have been presented, our attention will now turn to the synthesis algorithms.

3. Transformations

As a first step in algorithmic synthesis, the designer may want to transform or partition the design, either to improve its efficiency or to explore design alternatives. At the Algorithmic Level, the design can be behaviorally partitioned into concurrent processes and pipestages, or can be structurally partitioned across two or more chips or boards. At the Register-Transfer Level, the design can be transformed to allow the control step scheduler and data path allocator to obtain a more efficient implementation. This chapter describes some of the transformations implemented in the Workbench.

To allow the designer to explore Algorithmic Level design alternatives, a set of behavioral and structural transformations [Walker87, Walker88] has been defined, and implemented in the Workbench. Applied interactively by the designer before control step scheduling or data path allocation, each of these transformations affects the behavior of the design, the structure of the design, or both, while maintaining the inter-domain links.

Some of these transformations are similar to those used in optimizing compilers, and are used to transform the design from one based on a readable behavioral description to one with a more efficient control structure:

- Procedures can be expanded inline to eliminate unnecessary microsubroutine jumps.
- Individual operations can be moved into and out of the branches of decoding operations, often achieving a better packing of operations into control steps.
- Nested decoding operations can be combined, eliminating unnecessary levels of decoding.

Other transformations have a more extensive effect, dramatically changing the implementation of the design:

- Parts of the design can be transformed into concurrent processes, signifying that they are to be synthesized with separate controllers.
- The design can be pipelined, with the designer specifying both the number of stages and the placement of stage boundaries.
- The design can be structurally partitioned, defining chip or board boundaries.

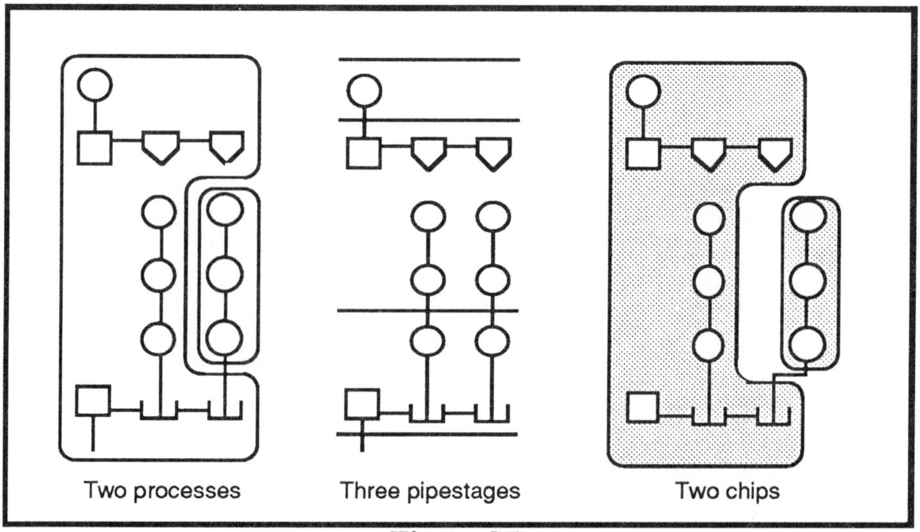

Figure 3-1
Alternatives in Behavioral and Structural Transformation

The last three alternatives are shown in Figure 3-1. In all three alternatives, an operation is performed, its result is decoded, and one of two SELECT branches is executed. However, in the leftmost alternative, the design has been behaviorally partitioned into two separate processes, resulting in a design with two smaller controllers instead of one larger controller. In the middle alternative, the design has been behaviorally partitioned into three pipestages. Each stage is a separate process, the three stages operate in lockstep, and data flows between the stages in a pipelined manner. With this partitioning, an increased instruction initiation rate should result. In the rightmost alternative, the design has been structurally partitioned across two chips, perhaps in response to physical constraints. Note that structural partitioning and behavioral partitioning are not mutually exclusive, as both might be applied to the same design. This chapter describes a set of transformations to support these three design alternatives.

Using these transformations, a designer can begin with a single "generic" behavioral description, transform the design to obtain a more efficient control structure, and behaviorally or structurally transform that design to explore further system-level design alternatives, such as concurrency or pipelining. As implemented in the System Architect's Workbench, these transformations can be interactively applied to the VT, the results can be analyzed (for example, by comparing the number of control steps or the amount of communication overhead), and additional transformations applied as necessary. The ability to consider alternative system-level organizations in a CAD environment based on a single behavioral description is a unique capability of the Workbench.

The next two sections of this chapter, *Vtbody Transformations* and *SELECT Transformations*, define transformations that can be used to transform a design from one based on a readable description to one with a more efficient control structure. The following two sections, *Adding Processes to the Workbench*, and *Process Creation*, define a transformation for creating processes. The next section, *Pipestage Creation* defines a transformation for creating pipestages. The next section, *Structural Transformations*, defines a set of transformations for directly manipulating the Algorithmic Level structural representation, and the final section, *Summary*, summarizes the chapter.

3.1. VTBODY TRANSFORMATIONS

As described in Chapter 2, when the ISPS behavioral description is mapped onto the VT, procedures and labeled blocks of operations are mapped onto VT subgraphs called *vtbodies*. CALL and LEAVE operations are used to transfer the flow of control between these vtbodies, similar to the CALL and RETURN operations in many programming languages.

The vtbody transformations allow the designer to remove and add CALLs between vtbodies by expanding vtbodies inline and forming new vtbodies. Since each CALL in the VT is implemented by a microsubroutine jump in the final design, these transformations give the designer control over the placement of the microsubroutine jumps.

3.1.1. Vtbody Inline Expansion

Vtbody Inline Expansion replaces a CALL operator with the contents of the called vtbody. In the Workbench, CALLs are not expanded if the vtbody being CALLed has more than one LEAVE operator, or if it RESTARTs, RESUMEs, or TERMINATEs itself; these restrictions avoid GOTO-style jumps to the interior of a vtbody (which are not allowed in the VT).

An automated version of this transformation is also available, that either expands inline all CALLs to the specified vtbody, or recursively expands inline all CALLS inside the specified vtbody. An example of the latter is shown in Figure 3-2, where all CALLs in vtbody V1:A are recursively expanded inline; the resulting VT is shown on the right side of the figure.

As implemented in the Workbench, a CALL operation performs a microsubroutine jump, and a LEAVE operation performs a microsubroutine return. Thus, if the VT in Figure 3-2 was synthesized before the transformation, at least three cycles would be required for its execution: one to execute the CALL and the instructions before it, one to

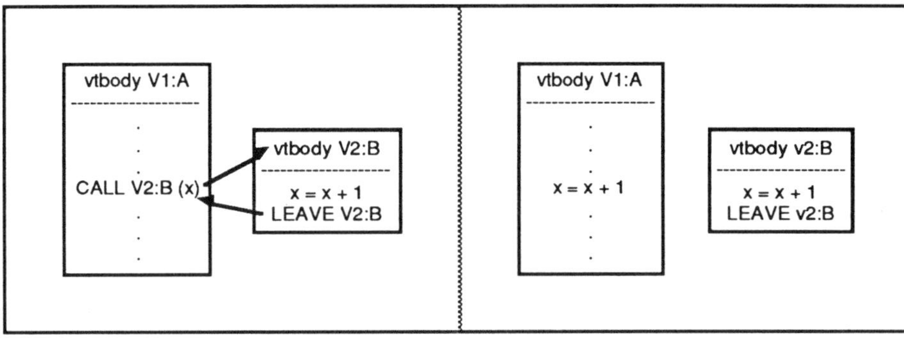

Figure 3-2. Before and After Vtbody Inline Expansion

execute the operations in the called vtbody, and one to execute the operations following the CALL. In contrast, after the transformation it is possible that all the remaining operations could be packed into a single cycle.

3.1.2. Dead Vtbody Elimination

Similar to dead code elimination in optimizing compilers, *Dead Vtbody Elimination* removes a vtbody that is not called by other vtbodies. An automated version of the transformation is also available that removes all vtbodies not called by others. For example, applying Dead Vtbody Elimination to the VT on the right side of Figure 3-2 would remove vtbody V2:B, which is no longer referenced by other vtbodies. This transformation is typically applied after Vtbody Inline Expansion.

3.1.3. Vtbody Formation

Vtbody Formation encapsulates a specified set of operators into a new vtbody, which is then CALLed from the original vtbody. This is the inverse of Vtbody Inline Expansion. An example of Vtbody Formation is shown in Figure 3-3, where an increment statement and its contents are encapsulated into a separate vtbody.

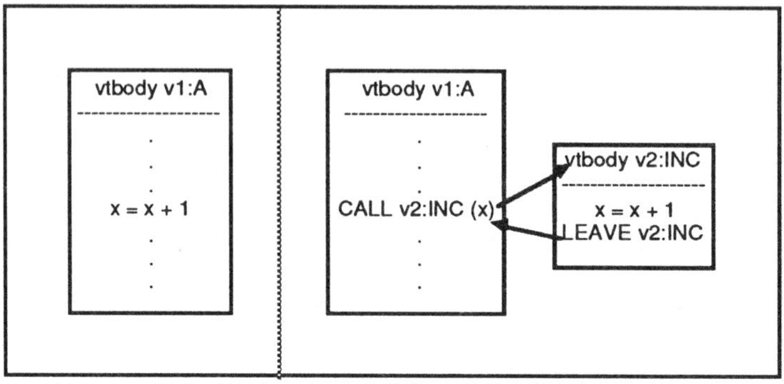

Figure 3-3. Before and After Vtbody Formation

3.2. SELECT TRANSFORMATIONS

As described in Chapter 2, in ISPS, IF operations are used for conditional branching, and DECODE operations are used for CASE-like decoding. Both of these operations are mapped onto the VT *SELECT* operation. This SELECT operation decodes a value (called a *selector*), and based on the decoded value chooses one of a set of alternate branches for execution; associated with each branch is a set of *activation values* that must contain the selector for the branch to be chosen. At any given time, only a single branch of the SELECT can be active.

The SELECT transformations allow the designer to interactively manipulate SELECTs and the operators surrounding them. Operators can be moved into and out of SELECTs, nested SELECTs can be combined, and SELECTs can be duplicated.

In the simplified VT diagrams in this section (Figures 3-4 through 3-12), VT operators are represented by circles and SELECTS are represented by squares. SELECT branches and their contents are represented by five-sided polygons and the operations below them. For example, Figure 3-4 illustrates a SELECT with three branches, each branch containing two operators, and an operator before and after the SELECT.

3.2.1. SELECT Motion

The SELECT Motion transformations are applied interactively to move operators into or out of SELECTs as shown in Figure 3-4. *SELECT Motion Down Into SELECT* moves an operator from above a SELECT into the top of all the SELECT's branches; *SELECT Motion Up Into SELECT* moves an operator from below a SELECT into the bottom of all the SELECT's branches. *SELECT Motion Down Out of SELECT* moves an operator from inside a SELECT to below the SELECT, and deletes duplicates of that operator in the other branches; *SELECT Motion Up Out of SELECT* is defined similarly, but moves an operator above the SELECT.

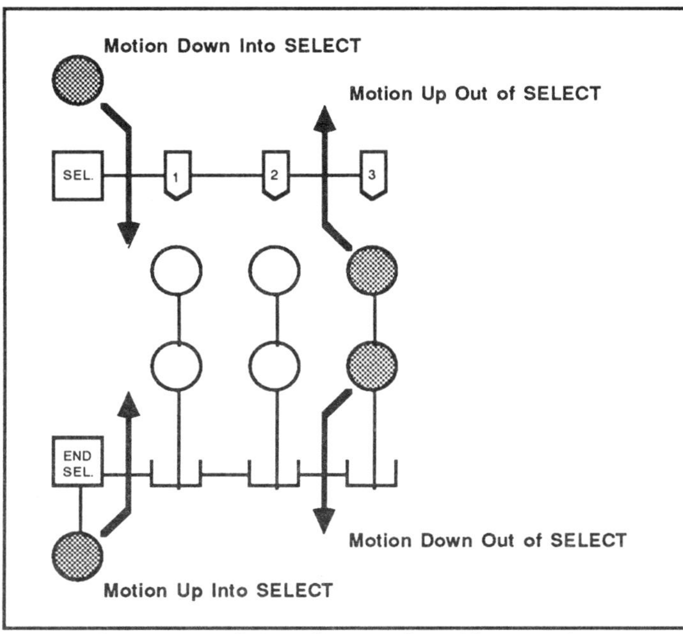

Figure 3-4. SELECT Motion Alternatives

An example of SELECT Motion Up Into SELECT is shown in Figure 3-5. In this example, the shaded operator is moved from below the SELECT into the bottom of each branch of the SELECT. Reforming the control structure in this manner might be useful to achieve a better control step packing, or to facilitate other transformations (e.g., the SELECT Combination transformation described in the next section).

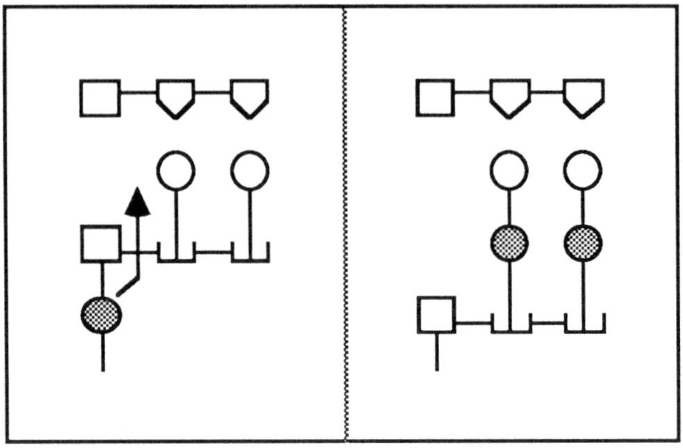

Figure 3-5
Before and After SELECT Motion Up Into SELECT

3.2.2. SELECT Combination

The *SELECT Combination* transformation is applied interactively to combine two nested SELECTs into one SELECT, replacing the branch of the outer SELECT that contains the inner SELECT with all the branches of the inner SELECT and updating the branch activation values accordingly. Thus, only one decoding operation will be performed, rather than two sequential decoding operations.

Two versions of SELECT Combination have been defined. In the first version, the selectors of the two SELECTs are concatenated to form a new selector, as shown Figure 3-6. In this figure, "<r>" indicates reading a bit field of value "r", and "@" represents concatenation of two values. In the second version, if the two selectors reference either the same value or subfields read from the same value, the full value is used as the resulting selector.

The first version, which concatenates the subfields, is more suitable for selectors with widely separated subfields to be decoded (for example, an instruction word with subfields separated by immediate operands). The second version, which uses the full value, is more suitable for selectors which are decoded completely, or for short selectors, where few

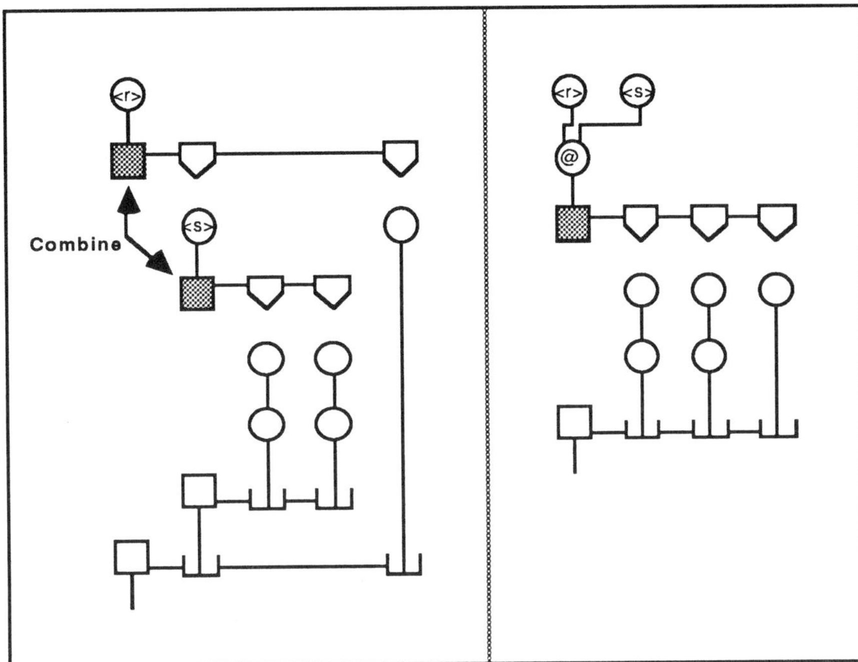

Figure 3-6. Before and After SELECT Combination

don't care values would result from the undecoded subfields. For example, given a value declared as <3:0>, combining a SELECT decoding bits <3:2> of that value with a SELECT decoding bits <1:0> of that same value should use the full-value version of SELECT Combination. In contrast, given a value declared as <31:0>, combining a SELECT decoding bits <31:30> of that value with a SELECT decoding bits <1:0> of that same value should use the concatenated-subfields version of the transformation, resulting in a 4-bit selector, rather than in a 32-bit selector with 28 bits of don't care values.

3.2.3. An Example of SELECT Motion and Combination

SELECT Motion and SELECT Combination are often used together to reorganize the instruction decoding in a processor. An example is shown in Figures 3-7 through 3-9. In Figure 3-7, the outer SELECT (1) decodes an opcode. The inner SELECT (2) then decodes the addressing mode and calculates the effective address (3 or 4). The operand is then fetched (5), the operation is executed (6), and the result is stored (7). For simplicity, the calculation of the destination effective address is not shown. As shown in the figure, these operations will require a minimum of 4 cycles to execute, assuming a microprogrammed controller, because the beginning and end of each SELECT branch forces a jump to a new code sequence.

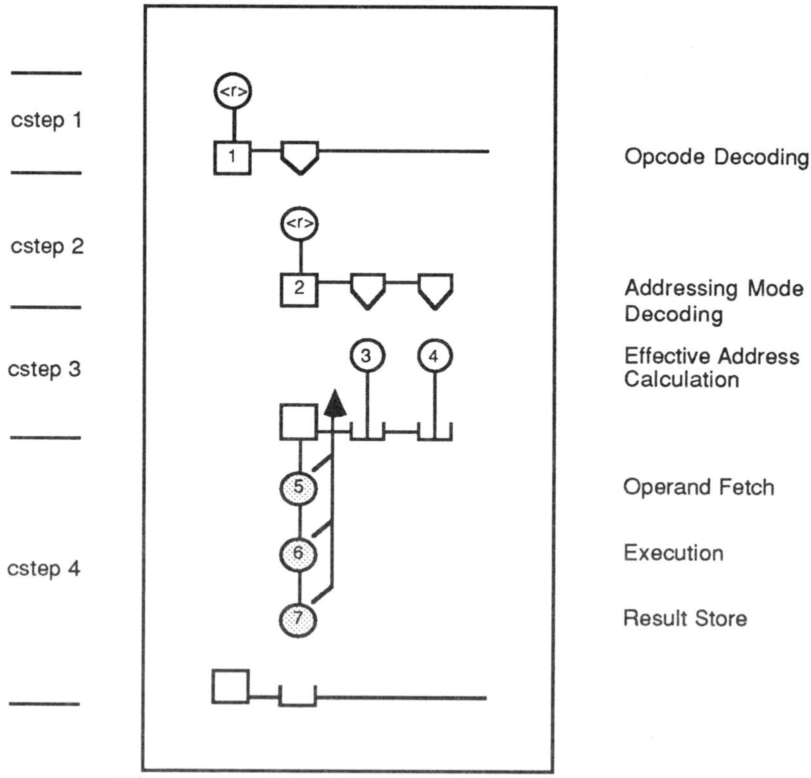

Figure 3-7. SELECT Motion Up Into SELECT

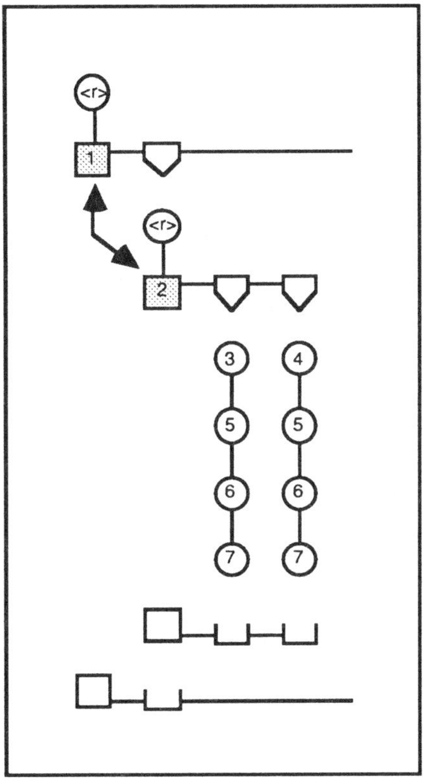

Figure 3-8
Before SELECT Combination

If the designer wants to combine the decoding and pack as many operations as possible into each control step, SELECT Motion Up Into SELECT can be applied to move the shaded operators below the inner SELECT up into that SELECT. The resulting VT is shown in Figure 3-8. SELECT Combination can then be applied, combining SELECTs 1 and 2, with the result shown in Figure 3-9. If all the operations in each branch can be combined into a single control step, the result would require only 2 cycles to execute – one for the SELECT and the operators before it, and one for the contents of the branch.

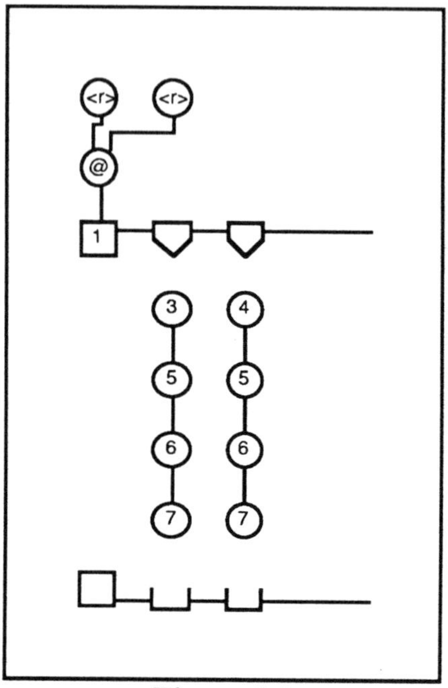

Figure 3-9
After SELECT Transformations

3.2.4. SELECT Duplication

The *SELECT Duplication* transformation is applied interactively to duplicate a SELECT, not including the operators in its branches; it places that copy into the VT immediately before the original SELECT. When the SELECTs are later implemented, it is assumed that the first SELECT in the series will perform the decoding operation once, and pass microcode dispatch addresses to the following SELECTs – this could be used, for instance, for functional branching [Tredennick80] or pipelining. An example of SELECT Duplication is shown in Figure 3-10, where the shaded SELECT is duplicated. Note that the operators in the branches of the shaded SELECT are not duplicated along with the SELECT.

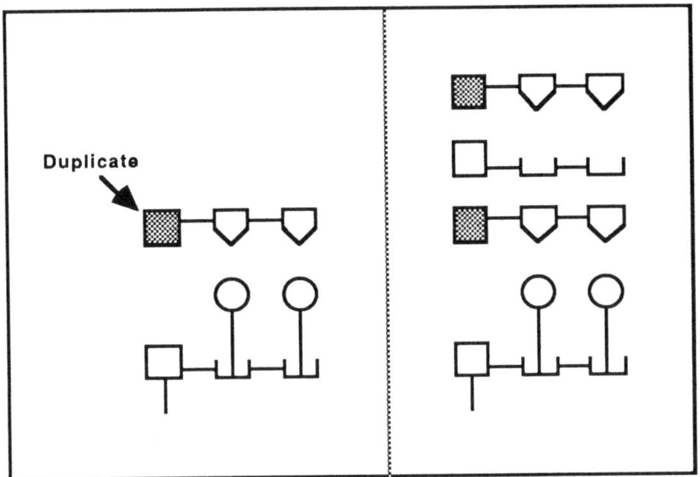

Figure 3-10. Before and After SELECT Duplication

SELECT Duplication is typically applied when creating a pipelined design. First, a set of "stages" are formed using SELECTs, and then these stages are encapsulated into vtbodies and transformed into pipestages. To form the SELECT "stages", SELECT Duplication is used to duplicate the instruction decoding SELECT for each of the other stages, and SELECT-to-SELECT Transfer (defined in the next section) is used to move operators into the appropriate stage.

3.2.5. SELECT-to-SELECT Transfer, SELECT Output Closure

The *SELECT-to-SELECT Transfer* transformation is applied interactively to move an operator from a branch in one SELECT to the end of the corresponding branch in the immediately preceding duplicate SELECT. An example of SELECT-to-SELECT Transfer is shown in Figure 3-11, where the shaded operator is moved to the corresponding branch in the previous SELECT; if each SELECT will later become a separate pipestage, this moves the shaded operator into an earlier stage.

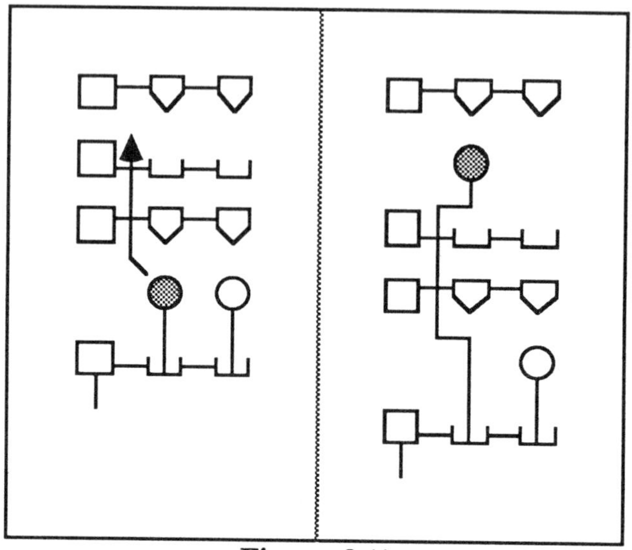

Figure 3-11
Before and After SELECT-to-SELECT Transfer

The *SELECT Output Closure* transformation examines the outputs of all operators inside the specified SELECT. If any of those outputs are used by operators following the SELECT, the VT is incorrectly formed. To correct the VT, the operator outputs being used improperly are connected to the output of the enclosing branch, an output is added to the SELECT, and that new SELECT output is connected to the operators that previously used the operator output. Outputs are also added to the other branches of the SELECT to maintain consistency (all branches must have the same number of outputs); these outputs become vtbody inputs added specifically for this purpose.

An example of SELECT Output Closure is shown in Figure 3-12, where the shaded operator causes its enclosing SELECT to be incorrectly formed. After the transformation, the output of that operator has been connected to the branch output, a new vtbody input has been added to the rightmost branch, and an output has been added to the SELECT.

SELECT Duplication, SELECT-to-SELECT Transfer, and SELECT Output Closure are often used together with the Process Creation and Pipestage Creation transformations described later to form pipestages.

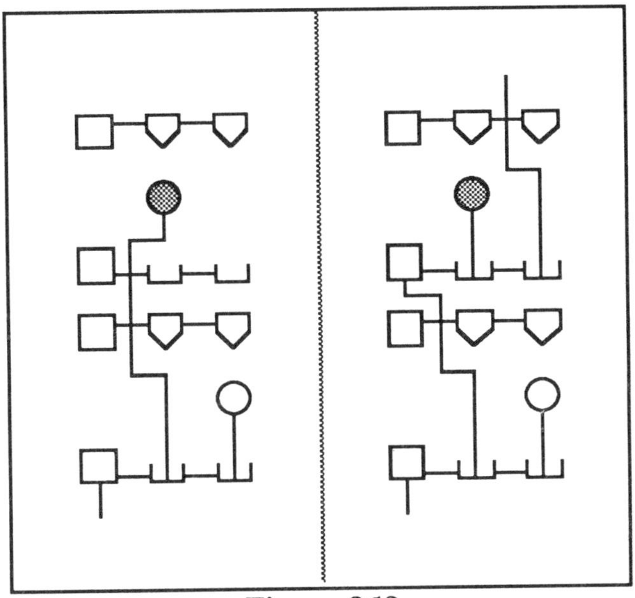

Figure 3-12
Before and After SELECT Output Closure

First, a set of "stages" are formed using SELECTS; these stages are then encapsulated into vtbodies and transformed into pipestages. The first step in forming these SELECT stages is to use SELECT Duplication to produce a copy of the instruction decoding (for a microprocessor design) SELECT for each of the other stages. SELECT-to-SELECT Transfer is then used to move operators into the duplicate stages. Once these stages are formed, each is encapsulated into a separate vtbody, transformed into a process, and then into a pipestage. A more complete example is given in Section 3.7.

3.3. ADDING PROCESSES TO THE WORKBENCH

This section describes methods for representing processes and inter-process communication mechanisms in ISPS and the Workbench, and two methods that could be used for implementing those processes and their communication in hardware. A transformation for creating these processes is described in the next section, along with its effect on the VT and on the final design.

3.3.1. Processes and Inter-Process Communication

In the field of computer science, processes, concurrency, and inter-process communication are well understood [Dijkstra68, Hoare78, Andrews83]. The execution of a set of sequential statements is called a *process*, and the parallel execution of two or more processes is referred to as *concurrent execution*. The transfer of data between two processes is called *inter-process communication*, and insuring that inter-process communication takes place at a mutually-agreed-upon time is called *synchronization*.

Several mechanisms are used to specify that two or more processes are to execute concurrently. One of the earliest, the *coroutine*, allows two or more processes to transfer control between themselves by means of the RESUME statement. Another early mechanism, the FORK statement, explicitly invoked a concurrent process, and then resynchronized on its completion by means of the JOIN statement. Finally, many modern languages simply label blocks of code with an explicit *process declaration*, with the understanding that all processes are to be executed concurrently.

Many mechanisms also exist for inter-process communication and synchronization. *Busy-waiting*, *semaphores*, *conditional critical regions*, *monitors*, and *path expressions* are all methods based on shared data – here the emphasis is on allowing only one process access to that data at a specific time. In contrast, methods based on *message passing* allow more explicit communication and synchronization – here two or more processes send explicit *messages* to exchange data, and must synchronize before that data exchange takes place.

In message passing, SEND and RECEIVE operations are used to synchronize, and to pass data via a specified communication channel. At one end of the message-passing spectrum, the SEND can refer to the specific process to which the information is sent, and the RECEIVE can refer to the specific process from which the information is received; this option is referred to as *direct naming*. At the other end of the spectrum, the SEND and RECEIVE can refer to global *mailboxes*. Since the mailbox is not directly tied to a particular process, this method is more general, and provides better support for client / server interaction, where there are multiple clients or servers.

3.3.2. Processes and Inter-Process Communication in ISPS

In the original ISPS language, processes are represented by attaching the PROCESS qualifier to a procedure or function. This qualifier indicates that the marked procedure or function is to be executed concurrently with the procedure that CALLs it, and is inactive otherwise. The qualifier CRITICAL can also be applied to a PROCESS, indicating that only one activation of the process may occur at a given time. It is assumed that the critical process has an *arbitration mechanism*, meaning that any attempt to activate a currently active critical process results in the blocking of the activating process.

Synchronization of ISPS processes occurs though process activation, and through the WAIT operation. A WAIT operation continuously evaluates an expression, and execution proceeds with the next statement when the value of the expression is not equal to 0. The last non-zero value of the expression is also returned by the WAIT.

ISPS processes transfer data in two ways. CALLing a process passes data to the process as formal parameters, and reading and writing shared carriers also passes data. These shared carriers can either be read directly, or read through the returned value of the WAIT statement.

This ISPS model of processes and inter-process communication is lacking in several aspects. First, communication through shared carriers makes it difficult to see where data transfer between two processes is taking place. Second, data transfer is inconsistent – some data is passed though the CALL, and other data is passed through shared carriers, although both methods may be implemented in hardware in the same way. Finally, communication (transfer of data) and synchronization (insuring that this data transfer takes place at a mutually-agreed-upon time) are separate and inconsistent: CALLs perform both synchronization and communication, WAITs perform synchronization and a limited amount of communication, and shared carriers perform only communication. Some of these concerns were also raised in [Parker79].

The following sections describe an (extended) ISPS / Workbench process model that explicitly represents data transfer between processes, uses only one set of operations for all data transfer, and merges synchronization and communication into one set of operations. This model is currently supported by Mill and the general synthesis path (CSTEP/EMUCS) of the System Architect's Workbench. The Process Creation transformation is available to create processes, and CSTEP and EMUCS can synthesize individual processes (although not yet considering the relationship between the processes).

3.3.3. Representing Processes in the Workbench

In the Workbench, a "process" is defined to be a ISPS procedure or VT vtbody marked as a process, plus the set of procedures or vtbodies recursively reachable through CALL and ENTER operations. When the distinction must be made between a vtbody marked as a process and the other vtbodies in the process, the former will be referred to as the "root vtbody" of the process, and the latter as the "child vtbodies", or "children", of the process; a similar distinction applies in ISPS. In the VT, the root vtbody of a process is marked by a flag in the vtbody header; children, which may be shared between processes, are unmarked. In ISPS, the root procedure is marked via the PROCESS qualifier.

ISPS / VT processes are always active. The last operator in a process is always a RESTART operator, and processes are never CALLed or ENTERed. A process can, however, CALL or ENTER child vtbodies.

Furthermore, if two or more ISPS or VT processes call a common child, that child will be part of both processes. If only one process is synthesized at a time, as is done in the current version of EMUCS, this should present no major problems, but if a tool were to attempt to synthesize the entire VT (i.e., all processes) at the same time, different allocation structures would have to be maintained for the common children for each process. Since maintaining multiple allocation structures might be difficult, one solution to this problem might be to define a "Duplicate Vtbody" transformation, which would allow each process to have its own copy of the common children. Another solution, using the existing transformations, would be to simply make the common children into a separate process.

3.3.4. Representing Inter-Process Communication in the Workbench

In the Workbench, ISPS and VT SEND and RECV message-passing operations are used to communicate between processes. These operations were chosen because they serve the dual purpose of communication (transfer of data) and synchronization (insuring that this transfer takes place at a mutually-agreed-upon time). As an implementation decision during synthesis, this synchronization may be insured through control step scheduling, or through explicit synchronization hardware; both of these options are discussed in the next section.

In ISPS and the VT, the input to a SEND operator is a set of values that are to be transmitted to other processes. RECV operators receive those values, blocking if no data is available, and pass them to other operators in the process through their output values. Both SEND and RECV operators contain qualifiers of the form "MESSG:<messgname>", which refer to a specific message. Since any given message might originate from one of many SEND operators, or be received by one of many RECV operators, these messages are assumed to be sent to and received from global *mailboxes* of that name.

Expressed in ISPS, the basic format of the SEND used in the workbench is:

```
SEND {messg: messgname} (input1, input2, ... inputN)
```

where *messgname* specifies the destination mailbox, and *input1* through *inputN* specify the values to be transmitted. It is assumed that the mailbox has a bounded buffer (of an unspecified length), and if the buffer is full, the SEND must *block*, else execution continues with the next statement.

Expressed in ISPS, the basic format of the RECV used in the workbench is:

```
RECV {messg: messgname} ( : output1, output2, ... outputN)
```

where *messgname* specifies the source mailbox, and *output1* through *outputN* specify the values to be received. If no message is available, it is assumed that the RECV will *block*, delaying execution until a message appears. Note the ":", which is used to indicate that the following values are operator outputs.

For convenience, non-blocking versions of SEND and RECV are also available: NBSEND and NBRECV. The syntax is the same as for the blocking versions, with the addition of a returned single-bit status value. If the NBSEND or NBRECV was successful, a TRUE value (1) is returned. If the buffer sent to by the NBSEND was full, or no data was available for the NBRECV to receive, a FALSE value (0) is returned.

The *messg* data structure is a C structure that represents the message mailbox between the SEND and RECV operators. Each SEND and RECV operator has a "MESSG" qualifier that names that messg structure, and each messg structure has an id string, a list of all SEND operators sending that message, and a list of all RECV operators receiving that message. Thus, given a SEND, a RECV, or a message, it is possible to find out which vtbody and SEND operators can send the message, and which vtbody and RECV operators can receive the message.

3.3.5. Synthesizing Inter-Process Communication

The ISPS / VT processes and inter-process communication mechanisms described in the previous sections could be synthesized in hardware in many different ways. The inter-process communication might be synchronous, or it might be asynchronous. Synchronization of the SENDs and RECVs might be insured through control step scheduling, or through explicit synchronization hardware and microcode. The former alternative shifts the burden to the synthesis software at design-time, whereas the latter may generate additional hardware or microcode, but both are perfectly valid implementations.

This section describes two possible synthesis alternatives for the synchronization of SENDs and RECVs in the Workbench; since the Workbench does not yet fully support the synthesis of processes, either could be adopted by the system as a valid implementation. Other synthesis alternatives may also be possible, and equally valid.

Synchronization through Control Step Scheduling. This section describes a method for synchronizing SENDs and RECVs through control step scheduling. A SEND transfers its data to a queue, register, or wire, and execution of the instruction following the SEND continues. A RECV reads the data from that queue, register, or wire, and execution continues with the instruction following the RECV. To insure that this communication is synchronized, it is the responsibility of the control step scheduler to schedule each process to insure that the SEND and RECV execute in lockstep, with their respective control steps being executed at the same time. Note that this may not always be possible.

An example of synchronization through control step scheduling is shown in Figure 3-13. Here the RECV in process Y can be scheduled into a control step that executes at the same time as a control step in process X containing its corresponding SEND, and the operators computing the inputs for the SEND and those using the results of the RECV can be scheduled into those control steps as well. In this

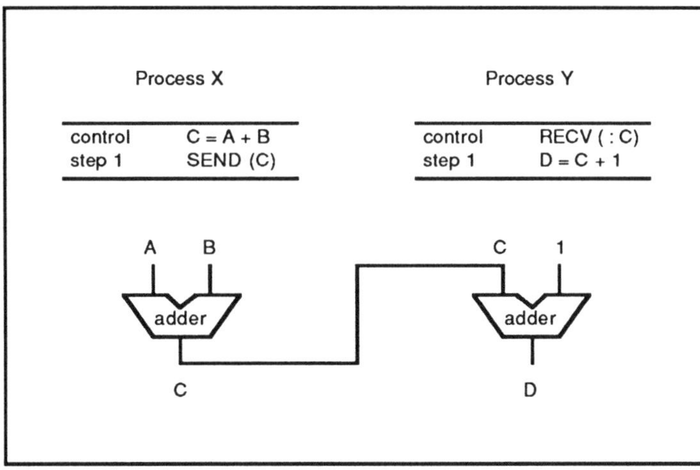

Figure 3-13
SEND / RECV Communication Implemented by a Wire

situation, the data transfer may be synthesized with a wire between the two data paths. However, if the operators computing the inputs for the SEND must be scheduled earlier, or those using the output of the RECV must be scheduled later, a register will be required to hold the values being transferred. This is shown in Figure 3-14, where the value RECVed is not used until two cycles later (due to operations not shown). In either case, no additional synchronization hardware or microcode is added to the design; insuring that the data being transferred is received directly or stored if necessary is the responsibility of the control step scheduler (and the register allocator).

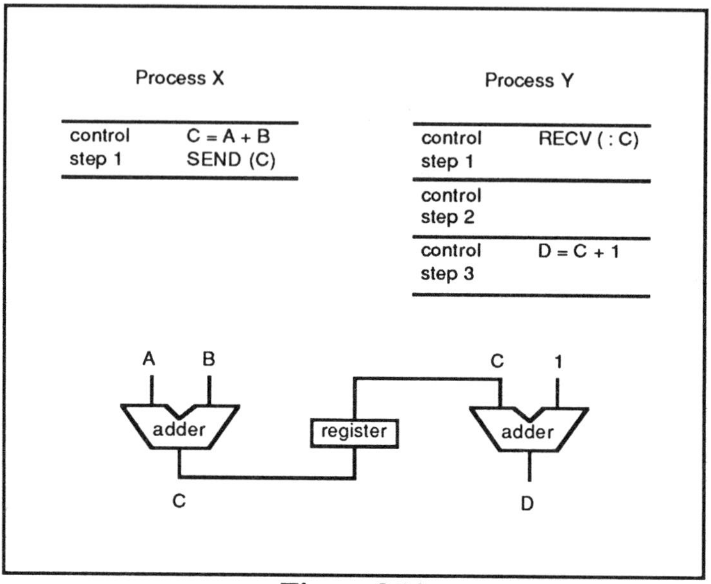

Figure 3-14
SEND / RECV Communication Implemented By a Register

To schedule operations in different processes so as to cause the SENDs and RECVs to execute in lockstep, the control step scheduler must consider not only the control steps in each process ("local" scheduling), but the relationship between the control steps of all processes ("global" scheduling). To achieve this lockstep execution, it may be necessary to insert delays into the control step schedule. This is shown in Figure 3-15, where a NO.OP (no operation) was inserted into the control step schedule of process Y to force the RECV to be executed at the same time as the corresponding SEND in process X.

It is not always possible to synchronize two processes through control step scheduling. If one of the processes contains a SELECT, and all branches of the SELECT do not require the same number of control steps, this technique is not particularly suitable, as all branches of the SELECT would have to be padded with delays to the length of the longest branch. However, if there is only one control path through each process, or all paths are the same length, synchronization through control step scheduling can avoid unnecessary hardware. For example, a three-stage pipeline, with each stage executing in exactly one cycle, could employ this method, and avoid the need for synchronizing hardware.

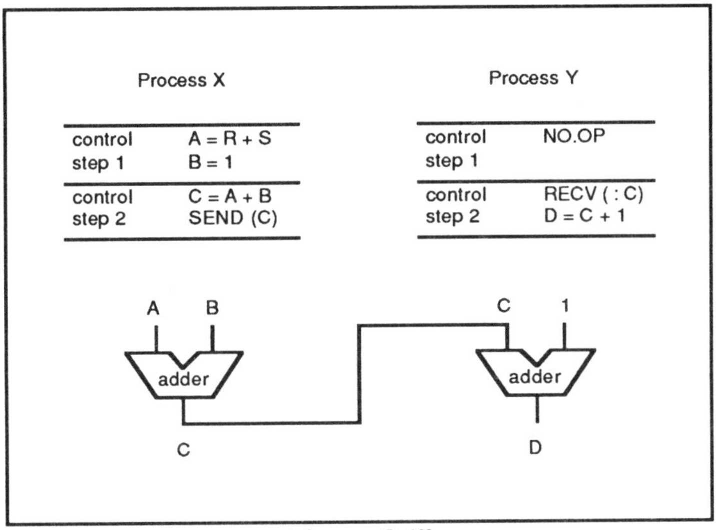

Figure 3-15
SEND / RECV Communication With Inserted NO.OP For Delay

Synchronization With Hardware Support. This section describes techniques employed to synchronize SENDs and RECVs with explicit hardware support. A SEND transfers its data to a queue, register, or wire, and execution of the next instruction (if any) continues. A RECV reads the data from that queue, register, or wire, and execution continues with the next instruction. If the data is not yet available, the RECV must block; the hardware or microcode implementation of the SEND and RECV must insure that this blocking occurs.

Several techniques are commonly employed for hardware synchronization, ranging from low-level primitives to higher-level mechanisms based on those primitives. A good survey of existing techniques is [Dubois88].

Lower-level techniques use atomic load and store operations, read-modify-write operations, or test-and-set operations to limit access to a section of code (called a *critical section*) to one process at a time. While one process is executing the critical section, it sets the *control variable* for that section to a particular value. If another process wants to also execute that section of code, it tests the control variable, finds out that the critical section is busy, and then blocks, usually repeatedly testing the control variable for access.

Higher-level mechanisms, such as message passing, can be implemented in microcode using these lower-level primitives for synchronization. For example, critical sections can be used to limit access to the code for reading and writing to a particular shared memory location, or to a mailbox. Depending on the complexity of the message-passing protocol used, additional microcode may also be necessary to test for a full/empty buffer, to manage global mailboxes, etc.

3.4. PROCESS CREATION

Previous sections of this chapter have described a set of vtbody and SELECT transformations. Those transformations, some of which are similar to the transformations used by optimizing compilers, are often used to produce a more efficient control structure for the design. In contrast, the transformations described in the remainder of this chapter have a more extensive effect, dramatically changing the implementation of the design. These transformations allow the designer to explore system-level design alternatives: forming concurrent processes, and pipelining the design's data-flow.

The first of these system-level transformations, the *Process Creation* transformation, creates a process from the specified vtbody and its children. It marks the vtbody as a process, replaces each CALL to the vtbody with a SEND / RECV pair of message-passing operators, replaces each LEAVE from the vtbody with another SEND / RECV pair, and RESTARTs the vtbody so that it will execute continuously. The new process will execute concurrently with the other processes, and will be synthesized with a separate data path and controller.

An example of Process Creation is shown in Figures 3-16 and 3-17. In this example, before the transformation, vtbody V1:A CALLs vtbody V2:B, passing it values x and y. After the transformation, the CALL has been replaced with a SEND operator, and a corresponding RECV operator has been added to the beginning of vtbody V2:B; values x and y are now passed via these message-passing operators. The LEAVE operator is replaced by a similar pair of SEND / RECV operators. After the transformation, vtbody V2:B is marked as a process, and is RESTARTed to execute continuously.

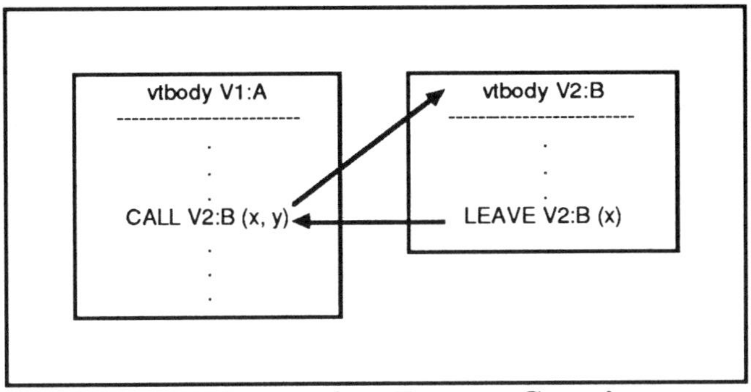

Figure 3-16. Before Process Creation

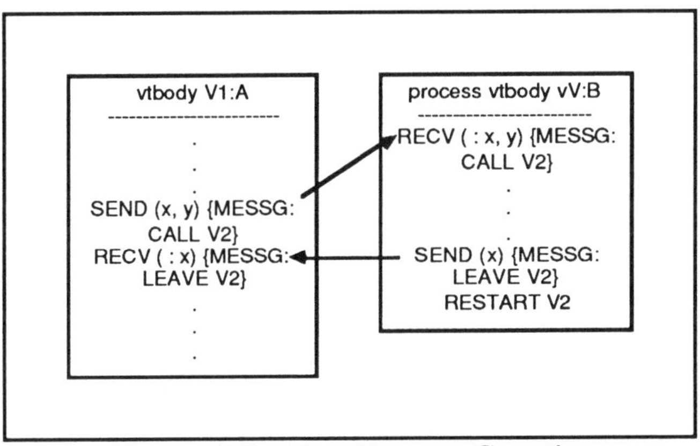

Figure 3-17. After Process Creation

Process Creation may be used to split a design in response to physical constraints, so that instead of producing a design with one large controller and one large data path that can not fit into the available chip area, two or more smaller controllers / data paths are produced, each of which can fit onto a single chip. This can result in potential for increased concurrency, in smaller data paths on each chip (but more total hardware), and in less total microcode. Each of these effects will be considered in the following sections.

3.4.1. Potential for Increased Concurrency

Process Creation also results in a potential for increased concurrency. Before the transformation, one vtbody CALLs another, waits for it to execute, and then receives any returned data. Scheduled in a similar manner after the transformation, one process SENDs data to another process, waits for it to execute, and then RECVs the returned data. However, if operators following the RECV are not data-dependent on the RECV, they can be scheduled between the SEND and RECV, so that the first process can be working while it waits for data to be returned from the second process. This concurrency does not result directly from Process Creation, but Process Creation allows this concurrency to be realized through control step scheduling. At this time, the CSTEP control step scheduler does not generate this kind of control schedule, although it could be modified to do so.

3.4.2. Less Hardware Per Process

One reason for applying Process Creation is to partition a design that is too big to fit onto a chip of a given size. When Process Creation is applied, other functional units (ALUs, adders, etc.) may be added to the design, and additional registers and multiplexors may be added due to the inter-process communication overhead. If this additional hardware is small enough relative to the total amount of hardware, each of the two processes will be smaller than the original process, and each resulting process may be small enough to fit onto a chip of that size. Experimental results (see Section 3.5) show that this is possible. Furthermore, since each of the resulting processes require less area than the original design, they may have reduced path lengths, a faster clock, and better performance.

3.4.3. Less Microcode Per Process

After Process Creation, the combined amount of microcode for the two resulting controllers may be smaller than the amount before the transformation. This can be shown as follows. Assume that before Process Creation is applied, the design contains m control steps, each corresponding to a microword. Further, assume that in each control step, s bits are required to specify the next control step (the next state), o bits are required to specify the microoperation to be performed (jump, jump to subroutine, continue, etc.), and d bits are required to control the data path. The total amount of microcode, μC, is then equal to the number of microwords multiplied by the width of each microword:

$$\mu C = m\ (s + o + d)$$

If the design is then split into two processes using Process Creation, extra control steps Δm might be added, extra bits Δs_1 and Δs_2 might be required in each process to specify the next control step, and extra bits Δd_1 and Δd_2 might be required to control new data path elements. Letting k represent the number of control steps in the first process, the total amount of microcode, μC, is equal to the sum of the amount of microcode in each process:

$$\mu C = \mu C_1 + \mu C_2$$

$$\mu C_1 = (k)\ (s_1 + \Delta s_1 + o_1 + d_1 + \Delta d_1)$$

$$\mu C_2 = (m + \Delta m - k)\ (s_2 + \Delta s_2 + o_2 + d_2 + \Delta d_2)$$

where

$$s = s_1 + s_2$$

$$o \geq o_1, o_2$$

$$d = d_1 + d_2$$

Extra control steps Δm may be added when two operations that previously executed in a single control step are split across the two processes. In many cases, this increase does not occur (see the example of Section 3.5). When it does occur, it is usually small. If this increase in the number of control steps can be outweighed by a decrease in the width of each microword, the total amount of microcode will decrease.

When Process Creation is used to split a design into one with two processes, each process will have a smaller number of control steps than the original design (although the total number of control steps in the design may increase). As the number of control steps per process decreases, the number of bits required in each process to specify the next state may also decrease. This relationship is given by:

$$s_1 = \lceil \log_2 k \rceil$$

$$s_2 = \lceil \log_2 (m - k) \rceil$$

If the number of control steps stays the same, a decrease in the number of bits required to specify the next state for each process results in a decrease in microcode. For example, if the original design has 32 control steps, 5 bits would be required to specify the next state, for a total of (32) (5) = 160 bits of microcode to specify the next state. If Process Creation were applied to split the design into one process containing 1 control step, and another process containing 31 control steps, the total amount of microcode would be (1) (1) + (31) (5) = 156 bits, a 2.5% savings; note that the savings occurs solely in the first process, which has a decrease in the number of bits required to specify the next state. This savings can be maximized by splitting the original design so that the two processes have approximately the same number of control steps. Returning to the example above, if the design was split into two processes, each containing 16 control steps, 4 bits would be required in each process to specify the next state, for a total of (2) (16) (4) = 128 bits of microcode, a 20% savings. Experimental results show a similar savings; see the example in Section 3.5.

The number of bits, d, required to control the data path elements, occasionally, but not always, decreases as well. Since the Workbench assumes that functional units are "free-running", no microcoded control is necessary, and changes in the number of functional units have no effect on d. Instead, the change in the number of bits to control the data path comes solely from the change in the number of registers and multiplexors. As the previous section explained, the number of registers and the number of multiplexor input bits per process may decrease, although the overall numbers may increase due to communication overhead. As the number of registers or multiplexors decreases, so do the number of bits required to control them. Continuing the example above, if before Process Creation was applied there were 32 control steps, 30 bits were required to control the multiplexors, and 10 bits to control the registers, a total of (32) (30 + 10) = 1280 bits of microcode would be required to control the data path. If Process Creation splits the design into two processes, each with less data path elements than the original design, but with more total data path elements, there can still be a microcode savings. For example, if after the transformation, each process contained 16 control steps, and each required 25 bits to control its multiplexors and 8 bits to control its registers, a total of (2) (16) (25 + 8) = 1056 bits of microcode would be required by both processes to control the data path, a savings of 17.5%. Section 3.5 shows an example, split into two processes at different points, with savings in the amount of microcode required to control the data path ranging up to 26.3%.

3.5. PIPESTAGE CREATION

The *Pipestage Creation* transformation, acting as a follow-on transformation to Process Creation, converts a set of processes to pipestages. This is done by removing RECV / SEND pairs from the intermediate processes so that the final stage can send its results directly back to the first stage, and by adding the new assumptions that the stages will execute in lockstep with all data transfers between stages occurring at the same time. Examples of Pipestage Creation are presented in the next section.

Pipestage Creation may be applied to any intermediate process in a chain of process vtbodies formed through Vtbody Formation and Pipestage Creation, such that the intermediate process vtbody, Vj, is preceded by vtbody or process vtbody Vi and followed by process vtbody Vk, and SENDs and RECVs the sets of values A, B, C, and D as shown in Figure 3-18. Vj receives the sets of values A, B, C, and D from the previous stage Vi, and sends the sets of values C and D to the next stage Vk. Vj receives the set of values C from the next stage Vk, and sends those values along with the set of values A back to the previous stage Vi.

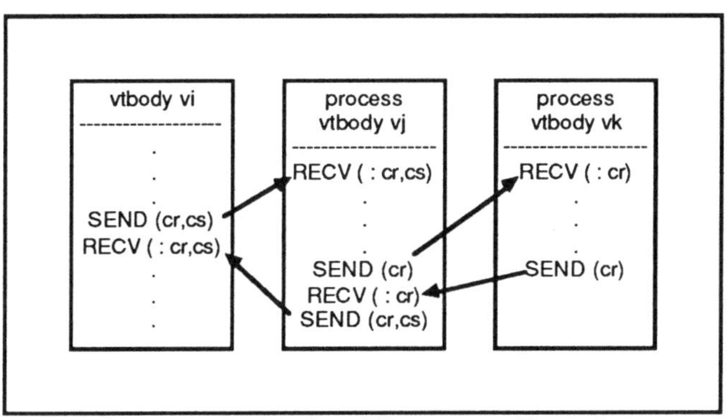

Figure 3-18. Before Pipestage Creation

Given an intermediate process of the form specified in the previous paragraph, the following occurs during Pipestage Creation:

1. The final RECV / SEND pair of operators in process Vj are removed.

2. The set of values A (that was sent back to the previous stage Vi, but that was not received from the next stage Vk) is sent to the next stage Vk, and from there to all successive stages until the final stage is reached, where it is sent directly back to the first stage.

   The result of these two steps is shown in Figure 3-19.

3. It is assumed that the pipestages will execute in lockstep, and all data transfers between the pipestages will occur at the same time. This assumption may be enforced either through control step scheduling, through added hardware or microcode, or by some other method.

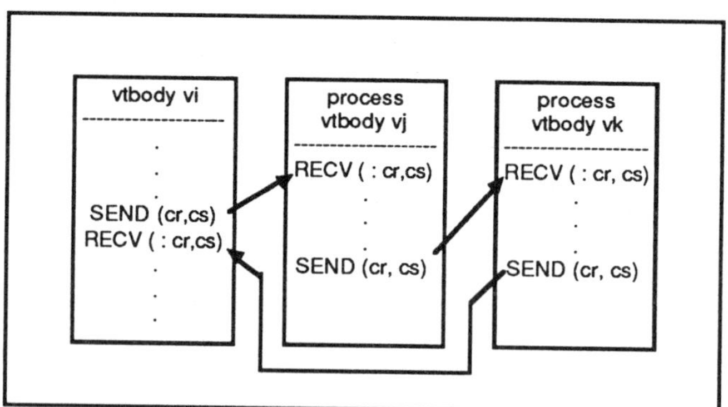

Figure 3-19. After Pipestage Creation

Together with the transformations described earlier, Pipestage Creation allows a designer to interactively pipeline the design's data-flow, choosing the stage boundaries and the operations to be placed in each stage. Thus a wide range of designs can be explored, ranging from designs with functionally decomposed stages (such as most CPUs) to designs with more balanced stages (such as pipelined multipliers or signal processing chips). This is in contrast to the Sehwa system [Park88], which automatically pipelines the design, but produces only the latter style of pipeline.

The preceding portion of Section 3.6 described a transformation for creating pipestages. Before this transformation can be applied, Process

Creation must be applied to form processes; before Process Creation can be applied, other transformations may be necessary. Choosing the "proper" set of transformations to apply can be difficult.

This section describes one path to follow in applying those transformations, in particular, transforming a "generic", unpipelined description of an instruction set processor to produce a pipelined design. The basic steps in this process are as follows:

1. Improve the control step schedule by expanding all vtbodies inline and by combining nested SELECTs.

2. Form "stages" using SELECT Duplication and SELECT-to-SELECT transfer.

3. Encapsulate the "stages" into vtbodies, then transform the vtbodies into processes and pipestages.

4. Analyze the results and iterate.

Improving the Control Step Schedule. Before forming pipestages, or even SELECT "stages", the VT should be transformed to make the control step scheduling as efficient as possible. For example, the original behavioral description of an instruction set processor may be very modular, possibly with individual procedures for each instruction, addressing mode, etc. It may also contain multiple levels of decoding, possibly decoding in sequence the instruction group, the instruction opcode, and the instruction addressing mode. While this formatting may make for a readable description, it may not be a very efficient control structure. With a microcoded controller, each of the calls to other routines forces a microsubroutine jump, and thus a new control step, which can possibly increase the number of control steps and thus the instruction execution time. Furthermore, sequentially decoding different fields of the same value can also increase the instruction execution time unnecessarily. With a few transformations, the control step scheduling can be greatly improved.

First, unnecessary microsubroutine calls can be eliminated by using Vtbody Inline Expansion to recursively expand all vtbodies inline, producing a single large vtbody if possible. Once this expansion has been performed, Dead Vtbody Elimination can be used to remove the vtbodies that are no longer referenced.

Second, nested levels of decoding can be combined using SELECT Combination. Before and during this combination, it may also be necessary to use various SELECT Motion commands to put the SELECTs into the proper form to be combined. See Section 3.2.3 for an example of using SELECT Motion and SELECT Combination.

Creating "Stages". Once the control step scheduling has been improved, the next step is to form a set of pipestages. This is done in two steps. First, SELECT "stages" are formed by duplicating the instruction decoding SELECTs using SELECT Duplication, and moving operators to the duplicates using SELECT-to-SELECT Transfer. Second, these stages are encapsulated into vtbodies, and processes and pipestages are created out of the vtbodies.

Before forming the vtbodies and final pipestages for an instruction set processor, the instruction decoding SELECT must be divided across the instruction branches to produce two or more SELECTs (one for each stage), each with the same branch structure, but with each SELECT representing a different "stage" in the processing of the instructions. For example, the designer may have decided to create a pipeline with IFETCH, DECODE, OPFETCH, and EXECUTE stages. In this case, SELECT Duplication would be used to duplicate the instruction decoding SELECT to produce a copy for the DECODE and OPFETCH stages, leaving the original for the EXECUTE stage (the IFETCH stage obviously does not need a copy). Once the SELECT has been duplicated, the designer can process the design on a branch by branch basis (for an instruction set processor, on an instruction by instruction and addressing mode by addressing mode basis) and use SELECT-to-SELECT Transfer to move operators to the duplicate SELECTs. Finally, SELECT Output Closure is used to clean up the resulting SELECTs.

Creating Processes and Pipestages. After the stages have been formed using SELECTs, the SELECTs are encapsulated into vtbodies, the vtbodies are transformed into processes, and the processes are used to form pipestages. First, the SELECTs are encapsulated into vtbodies using Vtbody Formation, and are formed so that each vtbody contains only a SELECT and its contents, a CALL to the next SELECT in the chain, and a LEAVE. Each vtbody is then transformed into a process using Process Creation, and the intermediate processes are transformed using Pipestage Creation. When these transformations are completed, the basic data flow is that of a pipelined system.

Analyzing the Results and Iterating. After the basic pipelined data flow is created, the user should schedule the control steps and simulate a representative set of instructions to estimate the performance of the design. Once control steps have been scheduled using CSTEP, the Workbench's Simulate Instruction command can be used to trace the execution of an instruction and determine the amount of time spent executing the instruction in each stage. However, since the CSTEP control step scheduler does not understand pipelined SELECTs, the results of the simulation must be processed slightly by hand to reflect this.

Once a representative set of instructions has simulated, it may be necessary to iterate further on the choice of stages and on the placement of operators into stages until acceptable results are achieved. However, since hazards, resource conflicts, etc. have not yet been taken into account, and the design has not yet been synthesized, this analysis should be regarded only as a first-order approximation.

3.6. STRUCTURAL TRANSFORMATIONS

In the Workbench's model of design representation, separate languages can be used to represent each of the three domains, and the domains can be coordinated and managed through inter-domain links. In the current implementation of the model, the Behavioral Domain is represented at the Algorithmic Level using ISPS and the VT, and at the Functional Block Level using the VT and the control step schedule. The Structural Domain is represented at both the Algorithmic and Functional Block Levels using a module / port oriented data structure. Although the Physical Domain is not currently implemented in the Workbench, it could also be described using this data structure.

Different tools are responsible for maintaining the different inter-domain links. CORAL, described in Chapter 9, builds and maintains the links between the ISPS description and the VT, and between the VT and the synthesized data path and control schedule. The transformations described in this chapter are responsible for maintaining the links between the VT and Algorithmic Level structural modules.

These transformations directly manipulate the Algorithmic Level structural representation and behavioral-structural links. They can be used to separate the behavior into two or more structural partitions, providing information for later synthesis tools. For example, the execute unit of a microprocessor can be separated from an instruction prefetch unit by encapsulating the behavioral description of each into a separate structural module; each may later be synthesized as a separate chip. Together with the APARTY architectural partitioner, these transformations can be used to separate the behavior into different structural modules, each of which contains one or more processes.

3.6.1. Structure Initialization

Structural Initialization creates an initial structural module with the specified name, and places all vtbodies in that module. An example of Structural Initialization is shown in Figure 3-20, where an initial structural module named "MAIN" is created.

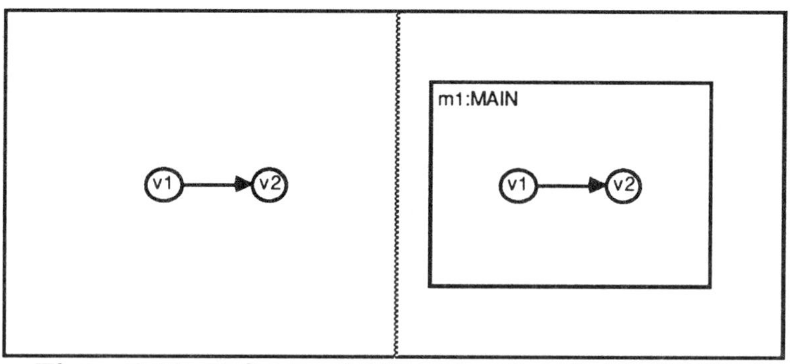

Figure 3-20. Before and After Structure Initialization

3.6.2. Module Decomposition

Module Decomposition creates a new module with the specified name, and adds it to the specified module as a submodule. An example of Module Decomposition is shown in Figure 3-21, where a new submodule is created in module m1:MAIN.

3.6.3. Module Removal

Module Removal removes a specified module. An example of Module Removal is shown in Figure 3-22, where module m3:B is removed.

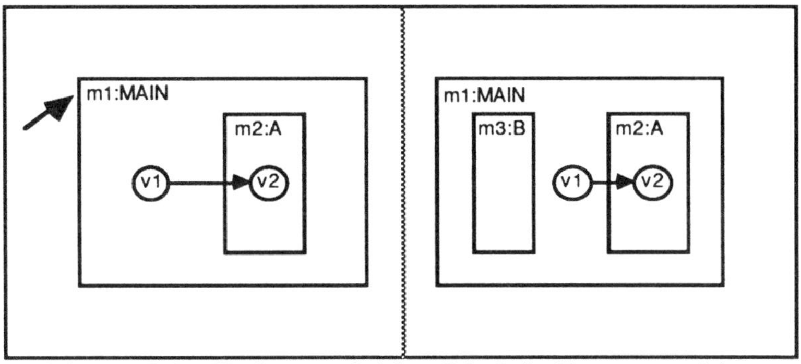

Figure 3-21. Before and After Module Decomposition

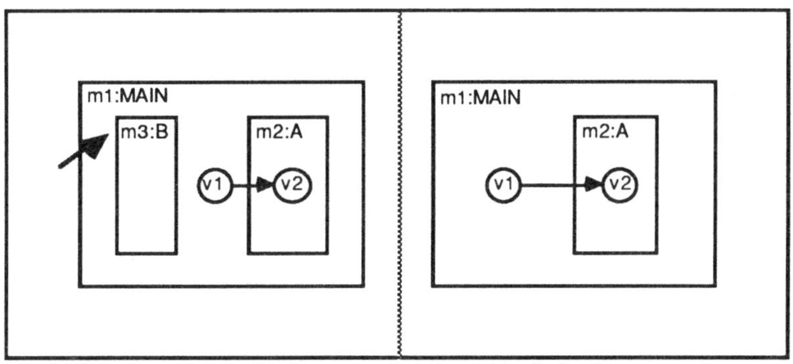

Figure 3-22. Before and After Module Removal

3.6.4. Vtbody Transfer (Between Modules)

Vtbody Transfer (Between Modules) moves the specified vtbody from its current module to the specified module. Any port connections between the current module and any other modules prior to the move are destroyed, and new port connections are added between the current module and other modules as required after the move. An example of Vtbody Transfer (Between Modules) is shown in Figure 3-23, where vtbody v1 is moved into module m2:A.

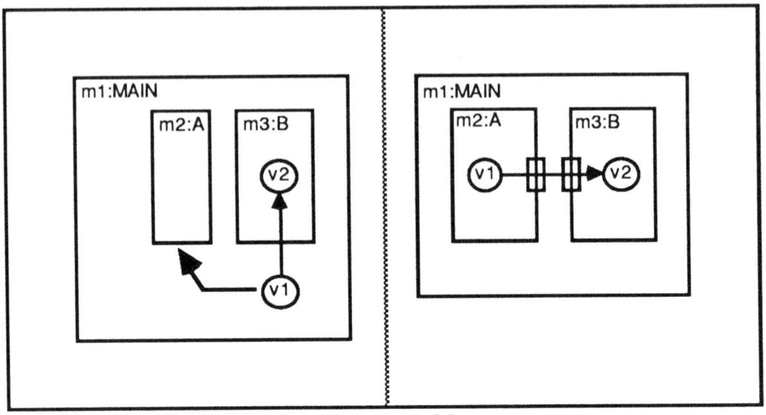

Figure 3-23
Before and After Vtbody Transfer (Between Modules)

3.7. SUMMARY

This chapter has described a set of behavioral and structural transformations that can be applied to the Value Trace. Lower-level transformations can be used to improve the efficiency of the design, and to facilitate higher-level transformations. Higher-level transformations can be used to explore Algorithmic Level design alternatives, including multiple processes and pipelines. Examples of this exploration are presented in Chapter 8.

4. Architectural Partitioning (APARTY)

Architectural partitioning is an early step in the Algorithmic Level synthesis process. It divides the behavior among high level physical units such as VLSI chips. There are several reasons for including this phase in the design process. One of the difficulties of synthesis is the lack of information in the behavioral description about physical aspects of the design that would aid in making decisions about the translation to the structural domain. This difficulty is compounded by the interdependency of design decisions. Architectural Partitioning addresses these problems by extracting high level structural information from a behavioral description. This information can be used in the synthesis process not only by giving preliminary clues about how synthesis should progress, but also by providing a global goal that can be used to maintain consistency across different design phases and allowing for a high level exploration of the design space early in the synthesis process.

Within the Workbench, architectural partitioning is accomplished by the APARTY tool which uses physical considerations to extract high level structural information from a behavioral description. The structural information extracted by APARTY is then used to guide synthesis in the Workbench. This structure affects both the scheduling of operators and the creation of the data path. Thus partitioning has a great effect on the outcome of the design. APARTY uses a novel multi-stage clustering technique that is based on an established clustering algorithm. This general clustering algorithm has been used in the past for data path design.

4.1. ARCHITECTURAL PARTITIONING

Architectural partitioning divides the behavior of an architecture onto separate physical units. Several criteria are important in deciding how the behavior should be divided. The most important architectural partitioning considerations are described below.

4.1.1. Area

Two important area considerations are addressed by partitioning: the size of the design relative to the physical units on which it is to be implemented, and the overall area of the design. Virtually any approach to partitioning reduces the amount of hardware that must be implemented in each partition. Good partitions may also reduce the global wire lengths by grouping the parts of the architecture that communicate heavily thereby reducing the interconnections needed to implement data dependencies between partitions and reducing the overall size of the design. The APARTY partitioner addresses both of these aspects of area considerations. Another aspect of the area consideration has to do with placement of modules on a physical device. Automatic layout tools have difficulty placing modules of vastly different sizes so it is desirable to have uniform partition sizes. APARTY attempts to restrict partition areas to a given window of acceptable sizes.

4.1.2. Performance

Performance may be improved through architectural partitioning by affecting wire length, operation level parallelism, and process level concurrency. Wire length impacts performance because of the time it takes to send signals over long distances, especially if the signals run off chip. Parallelism directly affects performance by shortening the control sequences. Concurrency affects performance by allowing multiple control sequences to execute simultaneously.

Two types of partitioning that address these performance issues are discussed below.

Data Partitioning. One method of improving performance is to keep together functions that make use of the same data, thereby improving the potential to exploit low level parallelism and decreasing the global wire length needed to interconnect functional units. Data partitioning groups operators with similar data usages in the same partition. This partitioning approach is best explained through an example.

Parallelism can be achieved through the data partitioning of the BTL310 architecture. The BTL310 is a digital signal processor having 6 partitions that are all implemented on a single chip: ROM, RAM, an address unit that updates memory pointers, a control unit to control operations in the other partitions, a data unit to execute the instructions, and an I/O unit to do serial data transmission and reception. Because data dependencies between the partitions are limited, register-transfers can execute in each partition simultaneously.

Control Partitioning. Long threads of control provide the potential for process level concurrency. In processor design, long threads of control correspond to machine instructions. Thus, instruction set partitioning is a type of control partitioning that attempts to keep all of the functionality necessary to execute an instruction together. Multiple instructions can then be grouped together to minimize the data interactions between instructions. The object of grouping instructions is to reduce wire lengths as well as the number of interconnecting wires between partitions.

Since complete instructions (or groups of complete instructions) are implemented on single partitions, the partitions have the potential for concurrency. Thus, each partition could be implemented as a separate process if data dependencies are satisfied. An additional benefit of this type of partitioning is that the data paths on each partition may be simplified because each partition implements only a subset of the total instruction set. A simpler data path may allow for a faster clock rate, and thus better performance for the design.

Three simple examples of this method of partitioning are the Intel 8086 [I8086], the NS16000, and the Zilog Z8000 [Faggin78], all of which have separate chips to implement floating point instructions. Each of these machines has a CPU chip to implement the non-floating point instruction set. The 8086 has the 8087 coprocessor chip, the NS16000 master chip has a 1600 slave chip, while the Z8000 has one or more

Extended Processing Units (EPUs) to execute floating point instructions.

In the 8086/8087 processor, an instruction intended for the non-floating point unit is ignored by the 8087. A floating point instruction is executed by the 8087, but the 8086 computes any addresses needed to execute the floating point instruction. Thus there is no concurrency in the design. In a similar fashion, the NS16000 master chip calls on its slave to perform the floating point instructions. The master waits for the slave to complete execution before continuing. The Z8000 configuration on the other hand allows instructions to execute in tandem. The Z8000 fetches all instructions. If it cannot execute an instruction, it issues a request for service from an EPU. While the EPU is executing the instruction, the CPU unit can continue fetching and executing instructions. This system has the potential for a great deal of high level concurrency, depending on the instruction mix.

The P1 described by Schmult [Schmult84] is a more complex example of instruction set partitioning. It has 3 partitions to implement the instruction execution cycles (one each for integer, floating point, and string instructions). This partition also has a separate chip to do all addressing including the execution of branch instructions. In this case, physically separating the instruction fetch from the instruction execution does not decrease performance because the partitions run concurrently. Thus the addressing unit is continually fetching new data to be fed to the execution units. The execution units can run concurrently as long as there are no data dependencies to prevent them from doing so.

Several instruction set partitions for the IBM System/370 are described in [Agnew82]. None of the methods took advantage of concurrency available at the instruction level. However, they represent a range of performance due to the way that the instruction set is partitioned in each case. The partitions are discussed in greater detail in Section 4.6.2.

4.1.3. Ease of Design

The ease with which a design can be developed can be broken down into two considerations at this level: the ease with which the partitions can be chosen, and the ease with which the partitions can be implemented. These factors are not of prime importance in a system where the design work is being done automatically, but examples of two partitioning schemes, control store partitioning and bit-slice partitioning, which address ease of choosing the partitions and ease of implementation of the partitions respectively, are mentioned here for completeness.

Control Store. The control store partitioning scheme is an example of a partitioning method for which partitions are easy to choose. In this partitioning method, the machine's data path is implemented on one chip while the microcode is stored on one or more separate chips. This is a relatively simple procedure because few decisions have to be made about where things are to be implemented. The control store can be divided onto as many chips as necessary, implying that the architecture can be extended easily by adding new control store partitions. Performance suffers in this scheme because the control store is in the critical path, and driving signals to and from the off-chip control store takes time. Also, this method generally requires a large number of connections between the partitions. This can be a problem in VLSI design, where the interconnect is limited by the number of pins available to each chip.

Examples of this type of partitioning are the LSI-11 [Dickhut80], the microEclipse [Druke80], and an implementation of the IBM/370 [Campbell82]. The LSI-11 is a chip set implementation of DEC's PDP11 architecture. There are three chips to the set: a data path chip, a control store chip, and a memory management chip. The microEclipse is a chip set implementation of Data General's Eclipse minicomputer. The partition consists of three types of chips: the CPU, one or more control store chips, and an I/O chip. By allowing a variable number of control store chips, the instruction set can be expanded. The IBM/370 chip set implementation consists of four chips: a processor unit, a storage addressing subsystem, a storage control unit, and a floating point execution unit, which is optional.

It is easy to determine how to implement a machine with control store partitioning but the implementation is bound to be slow. Therefore APARTY does not address this partitioning scheme.

Bit Slice. Another type of partitioning that addresses the ease of design consideration is bit-slice partitioning. The ease of implementing the partitions is the major consideration in bit-slice partitioning. In the bit-slice method, the data path is divided between several partitions, each one implementing some portion of the bits of the data path. The major advantage of bit-slice partitions is that since each of the data path partitions are the same, only one slice has to be designed. A disadvantage is that there is still an interruption of the critical path as information is transferred between partitions. The difficulty of designing according to this partitioning scheme is in creating a data path that can be divided appropriately; vertically sliced partitions are easy to choose after the data path has been chosen.

This type of partitioning is not addressed by APARTY because it is a problem of allocating the data path such that it can be divided vertically to produce the bit-slices, it is not a behavioral level partitioning problem.

4.1.4. Overview

The goal of APARTY is to be able to create both partitions with long threads of control and partitions that have reduced data interaction, and to choose these partitions such that they meet area restrictions. Partitioning for both control and data each has advantages. Long threads of control reduce the number of times that control must be passed between partitions and provide the potential for high level concurrency. The considerations that are addressed by APARTY reflect structure by reducing the need for control and data interaction between partitions. This provides high level guidelines for the synthesis tools that follow. Determining structure from behavior is difficult because of the lack of physical information available about the design early in the synthesis process. APARTY approaches the partitioning problem through a thorough evaluation of the behavioral description by means of a clustering algorithm.

The next sections will describe the algorithm used by APARTY to do the partitioning, discuss some of the specific criteria for doing the partitioning, and present an example of a design partitioned by APARTY. Additional partitioning results are given in Chapter 8.

4.2. PREVIOUS WORK: CLUSTERING

To choose architectural partitions, APARTY uses a unique multi-stage clustering technique. Multi-stage clustering, which is described in Section 4.3, is based on an established clustering algorithm that has been used in the past for data path design. The established algorithm is described in this section.

Clustering considers a set of objects, and groups them according to some measure of closeness. This is shown in Figure 4-1 where the objects A-E have been placed in space according to some measure for their closeness. Each pair of objects in the figure has a closeness value. The closest objects, A and B in this case, are clustered first and considered to be a single object for future clustering. Closenesses between the new object and the other objects in the space are estimated as either the maximum or the minimum of the closeness of the other object to A or B. The choice of maximum or minimum estimates is made according to the goal of the clustering. As clustering proceeds, new clusters are created by grouping two individual objects or by grouping an object or cluster with another cluster. Although the clustering is a binary process, several objects may be clustered together at the same level of the cluster tree if they have equal closeness measures. Thus clustering does not necessarily produce a binary cluster tree. The clustering process is iterated until a single cluster is produced and a hierarchical cluster tree has been formed.

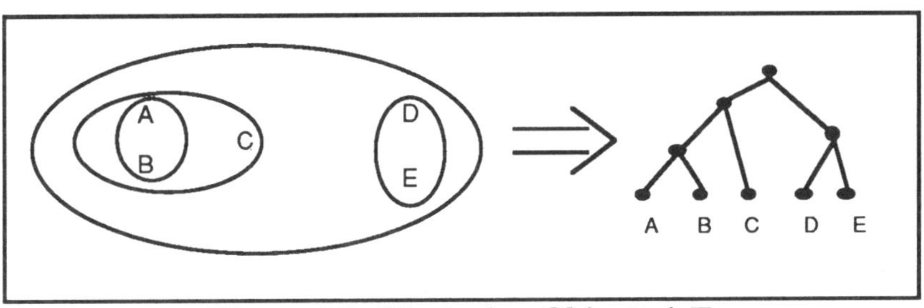

Figure 4-1. Clustering Objects A-E

The clustering process for several objects is shown in Figure 4-2. In the beginning, closenesses are measured and recorded in the lower triangular closeness matrix shown at the right of the figure, part *a*. A large closeness value indicates that the objects are very close, so the pair of objects with the largest closeness value is clustered first. Initially there are 3 pairs of equally close objects that have the maximum value for closeness: A, B; B, C; and D, E. Any one of these pairs might be chosen to make up the first cluster. Assuming that the first of equivalent closenesses in the matrix is always chosen by default, A and B will be clustered first. For this example, new closeness values are estimated as the maximum of the two previous closeness values. Thus, closeness values between the new cluster AB and the other objects are estimated as the maximum of their closenesses to A and B. The AB closenesses are stored in place of the A values, and the matrix is reduced by removing B from it.

In the next step, 9 is still the maximum closeness, and since AB, C is the first closeness of 9 that is found in the matrix, it is clustered next, as shown in part *c* of the figure, where C has been added to the AB cluster at the same level in the tree. The remainder of the clustering process is shown in parts *d-f* of Figure 4-2.

For more details on this clustering algorithm, see [Johnson67].

Clustering can be used to group data flow operators into meaningful groups. Clusters of operators may suggest functionality that should be combined in a single hardware unit. In order for the clustering to suggest meaningful hardware, the closeness between operators must be measured appropriately. When several criteria are used to measure a single closeness, weights must be used to assure the proper closeness contribution from each criterion. Several approaches to design automation using clustering are described by McFarland [McFarland83], Rajan [Rajan85], and Camposano [Camposano87a].

McFarland considers the similarity of functions, common data carriers, and low level parallelism to measure intra-procedural similarities between operators. The complete cluster tree of operators is created using this measure. A low cut through the tree produces groups of operators to be mapped onto register-transfer level hardware devices. A high cut produces clusters that correlate with high level partitions. McFarland has expanded his initial work in a program called BUD which creates a complete preliminary design from the clusters [McFarland86a]. The preliminary design is used to provide bottom-up information to DAA [Kowalski84]. The clusters provided by BUD are used to allocate processors and bind operators to them before turning over the synthesis to DAA.

Rajan used the same similarity parameters as McFarland in an early version of SUGAR, but used them to measure inter-procedural as well as intra-procedural relationships. The operator groups produced

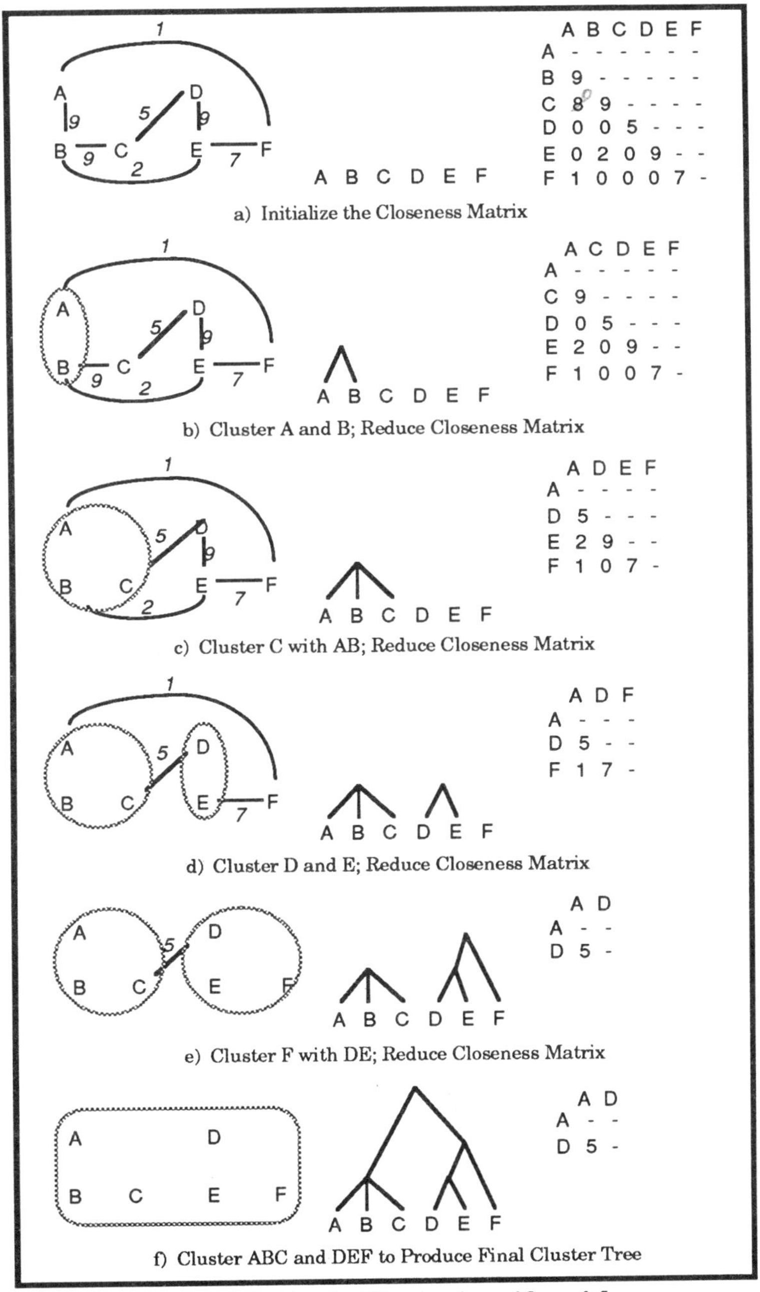

a) Initialize the Closeness Matrix

b) Cluster A and B; Reduce Closeness Matrix

c) Cluster C with AB; Reduce Closeness Matrix

d) Cluster D and E; Reduce Closeness Matrix

e) Cluster F with DE; Reduce Closeness Matrix

f) Cluster ABC and DEF to Produce Final Cluster Tree

Figure 4-2. Basic Clustering Algorithm

by the clustering were used as preliminary information to guide the allocation of register-transfer level hardware. SUGAR no longer uses clustering in its design process, as described in Chapter 7.

Like McFarland and Rajan, Camposano uses a single clustering stage which considers several criteria, but his clusters suggest partitions that are then used as input for logic synthesis. He considers similarity of functions and common data carriers in his clustering. The partitioning reduces the time spent in logic synthesis and may improve the synthesis results. The starting point and goal of this clustering are quite different from the behavioral clustering done by APARTY, BUD, and the early version of SUGAR.

Both BUD and the early version of SUGAR employ a single clustering stage geared a choosing groups of operators that should be implemented by individual functional units. McFarland may allow higher levels of the cluster hierarchy to represent higher level partitions. However, APARTY differs from both of these programs because it is specifically geared at high level partitions. These high level partitions may suggest some degree of low level parallelism as function units clusters do, but architectural partitions may also imply high level concurrency. Furthermore, architectural partitions can be used to guide not only the choice of functional units, but also the mapping of the data flow onto registers and busses, as is described in Section 4.5 and Chapter 6.

The biggest difficulty in choosing architectural partitions which is different from choosing functional units is the large groups of operators that must be considered. This difficulty points to several problems in using the basic clustering algorithm to choose high level partitions. First, basic clustering considers only individual objects rather than groups of objects in choosing clusters. But individual objects may not be able to reflect the flavor of the larger group. Furthermore, by considering only individual objects, closenesses between clusters can only be estimated as clustering proceeds, thus introducing larger and larger error as clusters increase in size. Finally, the difficulty of choosing weights for the various clustering criteria is only compounded in the larger problem. APARTY uses a novel multi-stage clustering algorithm which was formulated to address the special problems of architectural partitioning.

4.3. MULTI-STAGE CLUSTERING

Multi-stage clustering is a variation of the traditional clustering algorithm. Multiple stages make the clustering both easier to implement and more powerful, in that it can consider operators in groups rather than individually.

In multi-stage clustering there are several clustering processes, each one having its own closeness criteria. The clustering stages are consecutive and each stage builds on the results of the previous stage. In each stage, a complete tree of clusters is created. Then a cut line is chosen at some level of the tree, and each subtree below that cut line becomes an element to be considered for clustering in the next stage. The complete tree is built at each stage because this allows various cut lines to be evaluated and the best cut line chosen.

Each stage uses the clusters cut from the cluster tree created in the previous stage, as shown in Figure 4-3. In the figure, the boxes A-K have been clustered together to produce the tree in part *a* of the figure. The horizontal cut line produces 5 clusters indicated by the boxes *l-p*. In the next stage of clustering, shown in part *b* of the figure, the boxes *i-m* are the leaves of the new tree. Since the building blocks for the second clustering are the roots of the clusters created in the first stage, the second stage uses the combined features of all of the objects in each cluster when determining the clusters' closenesses.

There are several advantages of multi-stage clustering:

1. Decoupling the criteria makes them easier to manipulate. Changing the order of application of clustering criteria can change the style of partitions produced. Also, separating the criteria makes it easier to add new criteria to extend the applicability of multi-stage clustering.

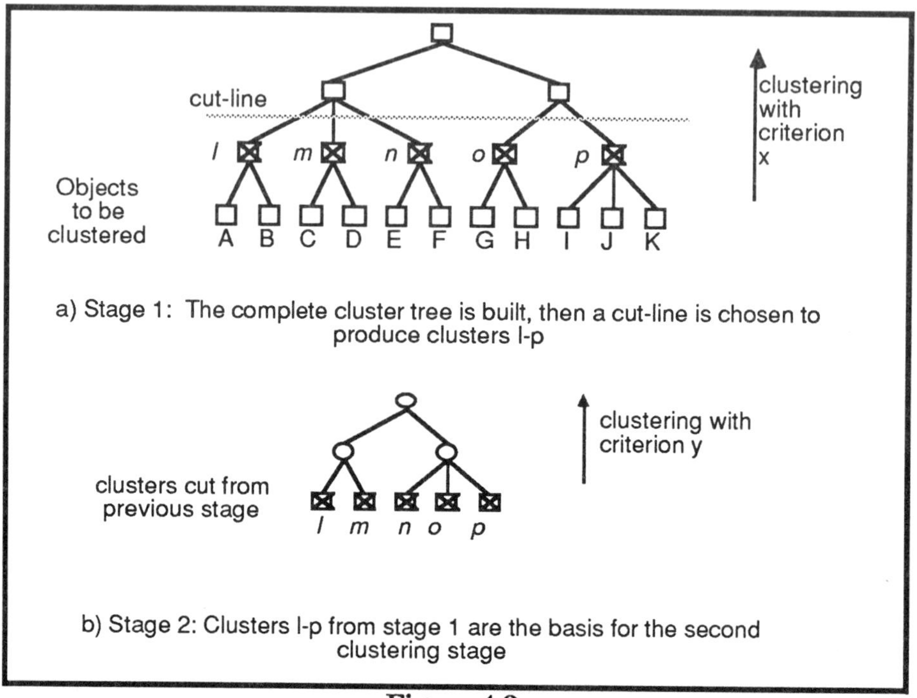

Figure 4-3
Two Clustering Stages, with the Clusters Cut from the First Stage Used as the Basis for the Second Stage

2. Multi-stage clustering limits the error introduced into the clustering process. As clustering proceeds, closenesses between objects are estimated each time a new cluster is formed. This estimation is an error factor that grows as clustering proceeds. However, by cutting through the cluster trees occasionally and recalculating closenesses, this error can be limited.

3. Multi-stage clustering allows operators to be considered as groups. This is particularly appropriate for architectural partitioning where large groups of operators are considered.

This is demonstrated in Figure 4-4. In part *a* of this figure, the closeness measure of the ten objects is indicated by the lines between the objects. Each line indicates a single closeness bond according to some criterion A. The sum of lines for each object is its total measure of closeness to all other objects. Since each object is equally close to 2 other objects, no one is tightly bound to any one other object. Thus the only cluster formed will contain all of the objects, as shown by the cluster tree at the right. Consider another stage of clustering using different criteria that might precede this and produce the groups **W-Z** as shown in part *b* of the figure. This stage uses a different criterion B, indicated by the dashed lines between the objects, to determine the clustering. In part *c* of the figure, the clusters produced by the B criterion clustering are clustered according to the original criterion A. The figure shows that groups **W** and **X** are closely bound (3 of their 5 total closeness bonds are common) and **Y** and **Z** are also closely related (5 of their 7 total closeness bonds are common). So this stage of the clustering results in objects *i-l* being grouped together, and *m-r* being grouped together. Without the previous stage of clustering that produced **W-Z**, all of the objects *i-r* would have been clustered together in a single trivial cluster, even though the same clustering criterion was used. Thus, very different groupings of the objects result from the 2 stage clustering scheme.

4.4. METHODOLOGY

APARTY's approach to the architectural partitioning problem uses the multi-stage clustering algorithm. The partitioning problem maps well onto this algorithm because multi-stage clustering can address the global view of the architecture with the various criteria that are important to partitioning. Several stages have been defined for partitioning use, each having different clustering criteria that address a particular aspect of partitioning: transfer of control, transfer of data, or hardware redundancy.

To create an architectural partitioning scheme for a design, the user chooses clustering stages according to the desired design style. In each cluster stage, a tree is built according to partitioning criteria that are used to measure the closeness between clusters. In the first stage, the VT operators are defined as the initial clusters that become the leaves of the tree that is built in this stage. In subsequent stages, the leaves of the trees are the clusters cut from the previous stage's trees. Cut-lines may be chosen by the user, allowing him to guide the clustering toward an appropriate partitioning scheme for the design. The choice of a cut-line affects the partitioning results insofar as it determines the number and size of clusters produced by the stage and the importance of the individual stage relative to the overall partitioning process. A low cut

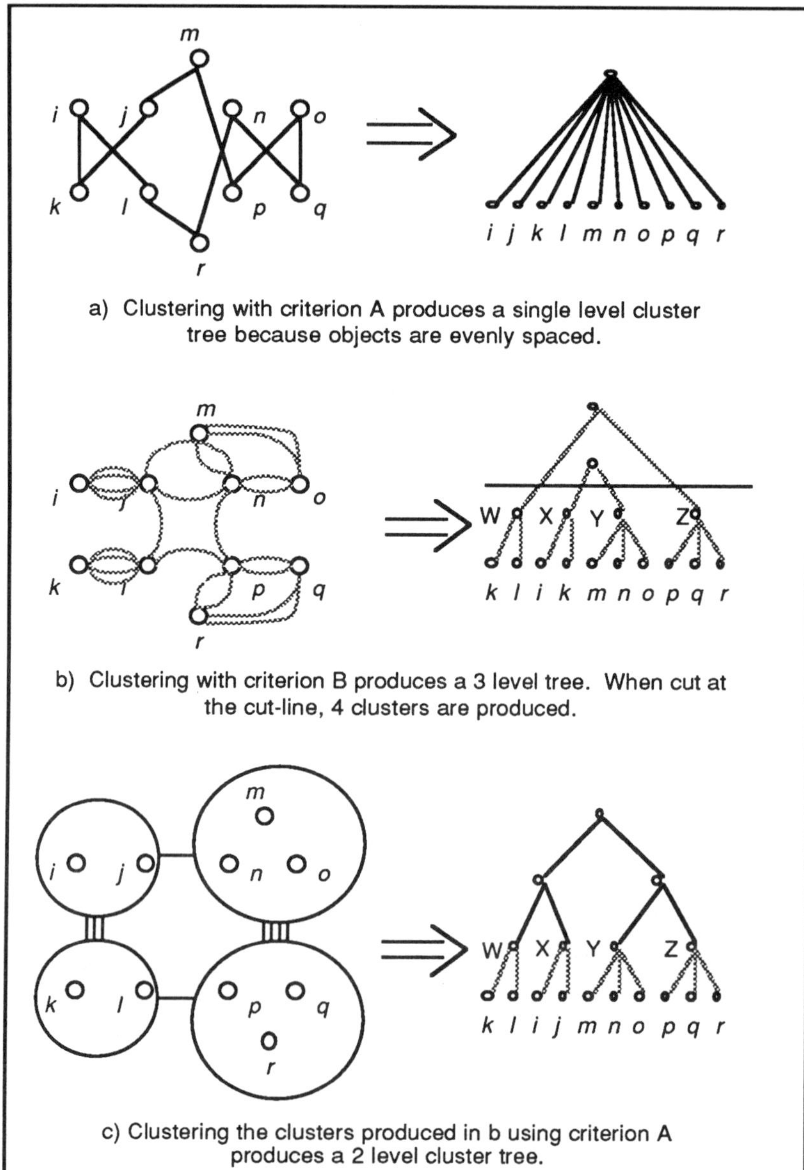

a) Clustering with criterion A produces a single level cluster tree because objects are evenly spaced.

b) Clustering with criterion B produces a 3 level tree. When cut at the cut-line, 4 clusters are produced.

c) Clustering the clusters produced in b using criterion A produces a 2 level cluster tree.

Figure 4-4
Different Groupings Result from Clustering with 1 and 2 Stages

in the tree causes a stage to have little importance because the majority of the clustering work done by the stage is ignored, while a high cut makes use of all the clustering done by the stage.

By including such considerations as data interconnect and control-flow in different stages, the various stages can be configured to target different partitioning styles which address different design needs. The ability to address different design criteria through partitioning is important because the partitions are used to provide a high level plan of the design to be followed by the synthesis tools. Thus partitioning has a great effect on the size and performance of the design.

The APARTY system includes 6 stages that can be combined to produce various styles of partitioning. Each stage attempts to produce clusters that optimize one partitioning criterion. The criteria addressed by the various stages are: the reduction of the number of times that control must be passed off between partitions, the reduction of the interconnect between partitions, performance in terms of schedule length, and the reduction of functional duplication between partitions. The specific considerations toward achieving these goals are: control flow, data flow, schedule length, data flow between vtbodies, control flow between vtbodies, and operator similarities. The user may choose which stages to run according to the type of partitioning desired for the design.

4.4.1. Control Clustering

The goal of the control flow clustering stage is to group together operators into long threads of control so as to reduce the number of times that control must be passed between clusters when the data flow is executed. APARTY provides a control clustering stage because it may be important to reduce the number of times that control is passed off between partitions for a particular design. The VT provides all of the information to find control paths within the data flow. Each control path represents a different way for information to be passed through the data flow, from the beginning of a vtbody to the end of it. The control clustering stage examines each control path, and by the appropriate choice of distance measures between the operators, attempts to unify the operators that share control paths. If it is to be used, control flow clustering is done first to guarantee the grouping of operators into threads of control. At one level, the threads of control may represent an architecture's instructions which can be clustered in turn to produce an instruction set partitioning scheme.

Definition. Control flow clustering is defined for individual data-flow procedures which are acyclic. Control flow between intra-procedural clusters is reduced by considering the probability that control will be passed from one operator to another in a single pass through the procedure. Groups of operators cannot be considered in this stage. More specifically, the control closeness between any two operators *a* and *b* is:

$$Cc_{a,b} = P(Op_b \mid Op_a)$$

where $P(Op_b \mid Op_a)$ specifies the probability that Op_b will be activated given that Op_a has been activated, and it is assumed that Op_b follows Op_a in the data flow. There are three possible configurations for two operators for this calculation. These configurations are demonstrated in Figure 4-5. The operators may be in different paths (e.g. operators 3 and 4 of Figure 4-5), so the probability of both of them being activated in a single pass through the procedure is 0. The operators may be in the same path with no branches between them (i.e. operators 4 and 5), giving a probability of 1 that both will be activated if it is known that one of them is activated. If, however, there is a branching configuration between the operators (i.e. operators 1 and 3), the probability that Op_b will be activated given that Op_a has been activated is equal to the probability that the branch that Op_b is in will be taken.

```
1. C = A + B;
2. if (C > 0) {
3. B = 1;
   }
  else {
4. B = 2;
5. A = A + 1;
   }
```

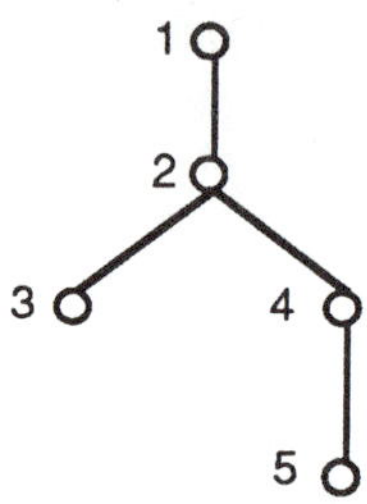

Figure 4-5. Control Flow Evaluation

Comparison. This clustering is different from the control flow criterion of McFarland and Rajan, (Camposano does not consider control) which considers the potential for low level parallelism between operators. The parallelism consideration employed by the other tools is used to separate operators that can be done in parallel so that they will be implemented by separate register-transfer level modules. Continuity of control flow is the consideration in APARTY, where clustering represents high level partitions rather than register transfer level hardware. APARTY is concerned with keeping operators that are in the same control path together so that control does not have to be passed off often between partitions.

4.4.2. Data Clustering

Data flow is another important consideration for partitioning. The data flow between operators is examined in terms of common data carriers. By combining operators that use the same data carriers, interconnect between partitions is reduced. Reduction of interconnect is important due to pin limitations between chips and routing congestion within chips that may make it impossible to complete the design. Reduction of interconnect also suggests that data dependencies between partitions are reduced. The reduction of data dependencies between partitions is necessary to performance because the greater the distance that information has to travel along a wire, the longer time it will take to arrive at its destination. It also aids parallelism. Since each partition will contain the hardware needed to implement its operators, the only restriction that might keep the partitions from running in parallel is the inter-partition data dependencies. Thus the implicit goal of this stage is to reduce the quantity of data that must be passed between clusters by reducing the number of values that flow between any clusters.

Definition. The explicit consideration of this stage is the simple ratio of the number of common values between the two operator clusters under consideration to the total number of values for each cluster of operators. Thus,

$$C_{D_{a,b}} = \frac{\text{Common}(a,b)}{V_a + V_b}$$

where *Common(a,b)* measures the number of common values between cluster *a* and cluster *b*, and V_a is the number of values flowing to and from cluster *a*.

A simple example of data similarities is shown in Figure 4-6. Since the PLUS and MINUS operators share data, even though the data is not aligned, these operators are considered to be very similar in terms of data usage. The PLUS and MINUS each have 2 of their three connections in common with the other. Thus their data similarity is:

$$C_{D\ PLUS,MINUS} = \frac{2+2}{3+3}$$

The AND operator shown in the figure has no data in common with the other operators and so would not be grouped with them.

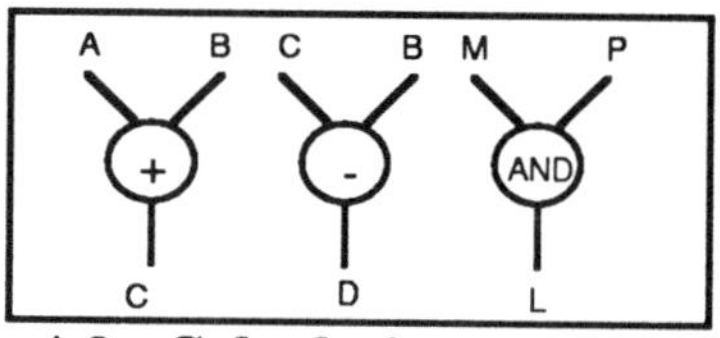

Figure 4-6. Calculating Data Closeness

Comparison. This is in contrast to other clustering algorithms which only count aligned operands because they are geared to choosing functional units rather than reducing overall interconnect between high level partitions. Furthermore, data flow clustering operates on clusters of any size, where each cluster contains one or more operators. This makes APARTY's data clustering different from the other clustering algorithms which only consider data similarities between individual operators.

4.4.3. Schedule Clustering

Basic data flow clustering ignores the issue of timing. Clustering operators together according to data similarities alone may lead to a poor schedule because operators that share hardware cannot be scheduled simultaneously. To take this into consideration, the scheduling criterion for clustering allows the potential for low level parallelism to be considered. However, since low level parallelism is a measure of distance rather than closeness, it is balanced with a measure of data similarities to get a closeness measure.

Definition. The schedule closeness calculation for two clusters *a* and *b* is:

$$C_{S_{a,b}} = C_{D_{a,b}} \times (1 - INC_{a,b})$$

where *INC* is the incompatibility for the two clusters being evaluated. The additive inverse of the incompatibility is used to measure closeness instead of distance between clusters.

Incompatibility is calculated by the following equation:

$$INC_{a,b} = \frac{\sum_{i=1}^{m} [I(a_i,b_1) \vee \ldots \vee I(a_i,b_n)] \times size(a_i)}{size(a) + size(b)} + \frac{\sum_{i=1}^{n} [I(a_1,b_i) \vee \ldots \vee I(a_m,b_i)] \times size(b_i)}{size(a) + size(b)}$$

where there are m operators in cluster a, and n operators in cluster b. The size function estimates the physical size of its argument according to the target technology for the design. $INC_{a,b}$ measures the incompatibility of all the operators in a as compared with the operators in b. Thus each operator in a is compared with each operator in b. If any two operators are incompatible, the penalty of putting them in the same partition is calculated as the size of the additional module that would be required to implement them without sacrificing schedule length. The sum of these penalties, normalized by the overall size of the clusters, measures the incompatibility for the clusters.

The incompatibility of a pair of operators is measured by

$$I(a_i,b_j) = h(a_i,b_j) \wedge g(a_i,b_j)$$

The function h determines if the operators reside in the same control step according to a previously determined ASAP schedule. g determines if the operators are of the same type. Thus if two operators might share hardware if placed in the same partition, but doing so would require more hardware in that partition to maintain the same schedule length, the incompatibility measure tends to push these operators into different partitions. At the same time, the data similarity measure tends to group operators that share data.

Comparison. This cluster stage is very similar to the clustering done by McFarland and Rajan, however multi-stage clustering allows groups of operators to be considered, making this more appropriate for the architectural partitioning application.

4.4.4. Control-Vtbody Clustering

In order to reduce the clustering possibilities in the early stages where there are many operators to consider, inter-vtbody clusters might not be considered by the control and data clustering stages. However, the inter-vtbody relationships cannot be ignored. The use of vtbodies within an architectural description is dependent on the programming style in which the description is written. In considering the vtbody interactions, inter-vtbody stages make an attempt to reduce the significance of the style in which the architecture is described, while combining appropriate function groups. To do this, vtbody clustering considers both from where the vtbodies are called and how often they are called. In this way, in a highly modular description where there are many vtbodies that are each called only a few times, the vtbodies will be grouped with the operators that call them, but commonly called vtbodies will not be grouped with a single calling sequence so that they can be shared easily. Thus a range of programming styles that make different use of vtbodies is permitted.

The objective of this vtbody clustering stage is to address control flow at a higher level than control flow clustering does. However, whereas control flow clustering considers acyclic graphs within vtbodies, control-vtbody clustering considers the possibly cyclic relationships between vtbodies. The general consideration remains the same: to reduce the number of times that control is passed between clusters.

Definition. Since this stage looks at a cyclic graph, it is appropriate to consider the probability that both cluster *a* and cluster *b* will be activated. So the closeness criterion is given as:

$$Cv_{a,b} = P(a \cap b)$$

$$= P(b) \times P(a \mid b)$$

where *P(a)* indicates the probability that cluster *a* is activated.

In the absence of a static evaluation of the data flow graph, the probability of control passing through all clusters in a given procedure is simplified to be equivalent. Therefore the probability of any one cluster being activated at any time is approximated by the product of the probability that the procedure it is in is will be activated and the number of clusters in that procedure. Thus if cluster *a* is in procedure *A*, the probability of cluster *a* being activated is estimated by

$$P(a) = P(A) \times \frac{1}{(\text{clusters in A})}$$

Assuming equal probabilities that any procedure will be activated,

$$P(A) = \frac{1}{(\text{total number of procedures})}$$

By the same assumption about the regularity of the control flow, the probability of control flowing through a cluster (*b*) given that control has passed through another (*a*) can be calculated by the ratio of the number of calls from *b* to *A* to the total number of calls to *A*. This measure is directional: if cluster *a* calls procedure *B*, but the converse is not true, *P(b | a)* is a positive value while *P(a | b)* is zero, which is incorrect. To account for this discrepancy of directionality, the closeness is measured from both directions and the highest measure is chosen as the one that is the correct measure of the probability of one procedure being activated given that another has been activated.

Comparison. Other clustering algorithms do not consider procedural relationships explicitly, although Rajan did measure inter-procedural as well as intra-procedural relationships between operators.

4.4.5. Data-Vtbody Clustering

While it is important to group vtbodies with the operators that call them for control purposes, it may also be appropriate to cluster operator groups that call the same vtbodies for data purposes. By clustering the calls together, they can share a communication mechanism which sets up variables in the same place in preparation for the call to the common vtbody.

Definition. The algorithm for doing this measures the common calls between two clusters relative to the uniqueness of the called vtbody, so that if two clusters call the same vtbody, and they are the only clusters that call that vtbody, they are more likely to be grouped together than if the called vtbody were called from many places. The equation for measuring Data-Vtbody closeness is given below.

$$Cw_{a,b} = \frac{\text{CommonCalls}(a,b)}{\sum_i \text{ExternCalls}(a, x_i) + \sum_j \text{ExternCalls}(b, x_j)}$$

where *CommonCalls(a,b)* counts the number of vtbody calls that are common between the clusters a and b, and *ExternCalls(a,* x_i*)* is the number of calls made to each vtbody x_i that is called from a.

Comparison. Again, other clustering algorithms do not consider procedural relationships explicitly.

4.4.6. Operator Clustering

The main goal of APARTY's architectural partitioning is to reduce both control and data interconnect between partitions. Reducing the overall hardware is a secondary goal, and therefore normally is considered only after control and data goals have been met by previous cluster stages. The means of reducing overall hardware is to cluster similar operators so that they can share data path modules. Two clusters are considered to be similar if they contain many of the same types of operators that cannot share the same control step. The operators' control steps are determined by running a scheduler previous to operator clustering to determine a preliminary schedule.

Definition. The operator closeness of two clusters *a* and *b* is given by:

$$Co_{a,b} = CMP_{a,b}$$

where compatibility for the two clusters being compared is measured by the following equation:

$$CMP_{a,b} = \frac{\sum_{i=1}^{m}[C(a_i,b_1) \vee \ldots \vee C(a_i,b_n)] \times size(a_i)}{size(a) + size(b)} + \frac{\sum_{i=1}^{n}[C(a_1,b_i) \vee \ldots \vee C(a_m,b_i)] \times size(b_i)}{size(a) + size(b)}$$

and

$$C(a_i,b_j) = f(a_i,b_j) \wedge g(a_i,b_j)$$

In these equations, the function *f* determines if the operators reside in different control steps, and *g* determines if the operators are of the same type. Operators are of the same type if they can be implemented together in the target technology. Thus the operator closeness between clusters indicates the degree to which it would be advantageous for the two clusters to be grouped together, and thus have the potential to share hardware.

Comparison. Operator similarities are used as criteria for clustering in other programs, but operator clustering in APARTY is different from the operator similarities measured by other clustering programs which consider the similarity of exactly 2 operators. APARTY's operator clustering does not consider individual pairs of operators, but rather compares operator clusters.

4.4.7. Summary

These clustering stages can be used consecutively to choose partitions for an architecture. The choice of which stages to run and the order in which to run them determines the style of partitions that will be formed, and is decided by the user. For example, the user may choose to apply the data and operator stages to produce data partitions, which will reduce the amount of data that flows between the partitions. An instruction set partitioning scheme will be chosen for an instruction set architecture if control clustering is used to group operators into threads of control that are the instructions. After that, data clustering can be applied to group the instructions appropriately, and the vtbody clustering stages and operator clustering can be applied as appropriate for the behavioral description. For more details, see [Lagnese89a].

The partitioning scheme chosen for a design is very important because partitioning determines the high level features of the design, which in turn affect the lower level features.

4.5. GUIDING OTHER SYNTHESIS TOOLS

The high level structural information provided by partitioning can be used to guide the data path synthesis process by providing a global goal for the design, allowing the tools to act in a more well informed and cohesive manner. Partitioning can be used to guide control step scheduling, data path allocation, the choosing of busses, and layout, because it provides preliminary clues about the physical aspects of the design. Partitioning reduces design considerations in the lower design levels by breaking the design problem into smaller parts. The high level structural information provided by the partitioner is more powerful than the information provided by other clustering algorithms because it can guide the choice of not only functional units, but also registers, interconnect, and layout of the hardware modules. By providing a global goal that each of the tools can work toward, even though they are independent, the tools can work toward a common goal and thus act in a more cohesive manner.

Partitioning information can guide control step allocation by providing a preliminary indication of the hardware that will be used to design the data path. Specifically, the partitions suggest a lower limit on the hardware that can be used for the design, because operators in different partitions cannot share hardware. Therefore, if PLUS operations are present in two partitions, the design will require at least two adders, and this information can be used in determining the appropriate level of parallelism in the control schedule.

Data path allocation can also make use of partitioning information. Since operations from different partitions cannot share hardware, a data path allocator can make use of the minimum hardware information that the control step allocator may take advantage of, but can also use the partitioning information to guide the mapping of data flow operators to hardware.

Data path allocation includes the allocation of busses. Choosing busses is a complex problem. Very long busses require large drivers and the time for a signal to travel the length of the bus increases with bus length. However, short busses may not justify the area expenditure required by the tri-state devices at the bus inputs. In the Workbench general synthesis path, busses are restricted to running between partitions, which implies long wires to justify the area expenditure, and only inter-partition signals may be busses to reduce the number of bus connections, and thus the size of the drivers required.

In the case of the Workbench, three tools are involved in the synthesis process: CSTEP is used to create a control step schedule, EMUCS allocates hardware and binds the data flow objects to that hardware according to the schedule produced by CSTEP, and Busser is used to choose busses for the data path produced by EMUCS. The specific use of partitioning information by these three tools is described in the CSTEP and EMUCS chapters.

4.6. A PARTITIONING EXAMPLE

Partitioning results are given here for the IBM System/370. Other partitioning results are presented in Chapter RESULTS.

The complete System/370 architecture is described in the IBM System/370 Principles of Operation [Principles80]. The instruction set is made up of general instructions which use the general register set, floating point instructions which use the floating point register set, decimal instructions which always operate on data in memory, and control and input/output operations which include all privileged instructions. A subset of the instruction set for this machine is described in its ISPS: most of the general, floating point, and control instructions, and approximately half of the I/O instruction are described while none of the decimal instructions are described in the ISPS. The ISPS description is quite modular, with each instruction defined as a separate procedure (each procedure is translated into a vtbody). There are also a number of auxiliary procedures that describe such functions as reading from and writing to memory.

Behavioral transformations were run on the System/370 VT to eliminate unnecessary operators, combine SELECT operators where possible, and inline expand any vtbodies that are only called once. This resulted in one large vtbody containing the gist of the instructions, and a number of smaller vtbodies containing auxiliary functions.

A research effort at IBM studied methods of partitioning the System/370 [Agnew82]. Instructions set partitioning was chosen as the best general method of partitioning primarily because of the reduced number of control breaks necessary with that partitioning scheme, and because it allows the partitions to be implemented on off-the-shelf microprocessors (Motorola's 68000). Seven instruction set partitioning schemes are described in the article that summarizes this research. APARTY's ability to address the various schemes is described below.

Of these seven partitioning schemes, four cannot be addressed by APARTY. Two of the four schemes require a core of operations to be implemented in hardware so that all instructions can be executed by software using that hardware. The problem of defining a sufficient core of operations requires that a minimal subset be found such that all instructions can be executed on the hardware. This is beyond the scope of APARTY. The remaining two schemes that cannot be addressed by APARTY use a combination of instruction set and control store partitioning. As described in Section 4.1.3, control store partitioning is uninteresting from an automated-design point of view because it is simple to design, but quite slow, so APARTY does not attempt to create this type of partitioning scheme.

Among the three remaining partitioning schemes for the System/370, the partitions can be described in terms of 5 functional groups: the instruction fetch cycle for all instructions; the execution cycles for fixed point, load/store and branch instructions; the execution cycles for floating point instructions; the execution cycles for privileged instructions; and the execution cycles for decimal instructions. These groups are combined in various ways to make up the three schemes, each of which has various advantages in terms of total area consumed, ease of design, and performance. A comparison of the results of partitioning using APARTY versus these instruction groups will be given in the following sections.

The System/370 description was run through four stages of the partitioner. First, control clustering was used to group operations into separate instructions. Next, data clustering was used to build groups of instructions that share data carriers. For example, this stage grouped all of the floating point instructions together because they use the same floating point carriers. Up to this point, each vtbody was considered separately for faster processing. In the third stage, vtbodies were grouped together according to the way they interact. Thus, the auxiliary floating point vtbodies that describe such functions as floating point addition and multiplication were grouped with the floating point instructions in this stage. In the final stage, clusters that made similar use of vtbodies were grouped together, so that in this stage, instructions such as the privileged instructions that make exclusive calls to a priority checking vtbody were grouped together. The final results of the partitioning are shown in Figure 4-7.

Figure 4-7 shows 3 of the 4 instruction groups mentioned in the article. Decimal instructions are absent because they are not included in the ISPS description. Twenty three "other" instructions are also represented. These are the instructions from the ISPS that do not clearly fall into any of the specified categories. The figure shows that there is a good correlation between the suggested partitions and the partitions chosen by APARTY. Of the 65 general instructions, APARTY grouped 59. Among the privileged instructions, APARTY failed to group 9, but of these, 3 were improperly defined in the ISPS, and therefore do not appear to be privileged instructions. The floating point partition chosen by APARTY is very clean. It contains exactly the floating point instructions described in the ISPS. Twenty-five instructions were not clustered. Most of these were "other" instructions.

This result indicates that APARTY is capable of addressing instruction set partitions, and makes reasonable choices of instruction groups in order to reduce data dependencies between the instruction subsets.

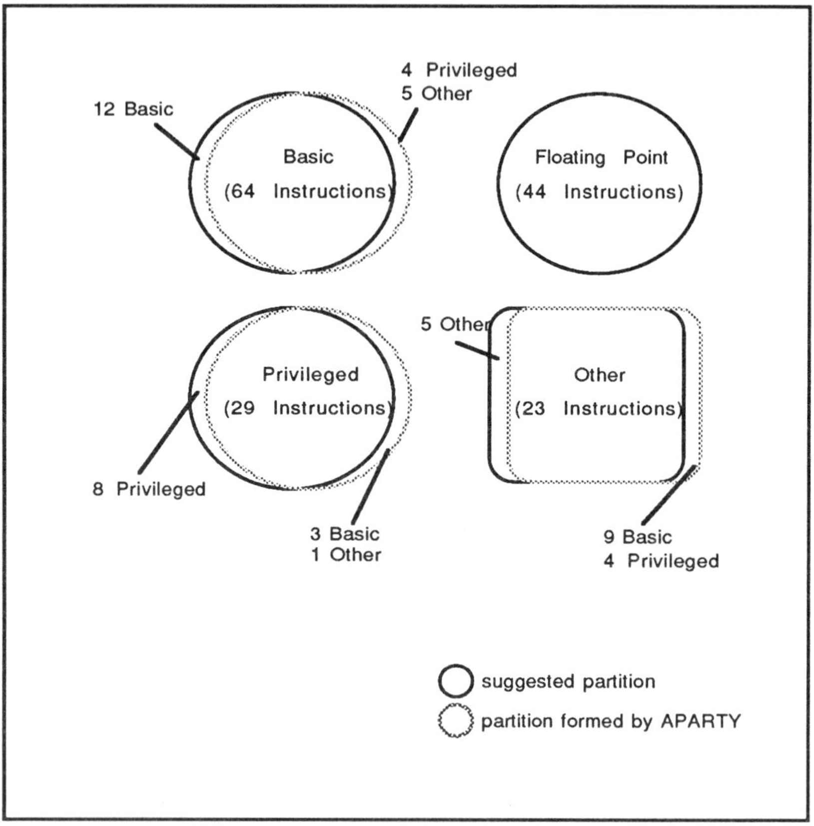

Figure 4-7. Partitions for the 370

4.7. SUMMARY

The multi-stage clustering algorithm used by APARTY is a new algorithm that is tuned to high level partitioning. The interchangeability of the stages makes this algorithm general enough to be able to address a wide range of both architectures and partitioning schemes, as will be demonstrated in Chapter 8. The partitions that are chosen provide high level structural information that can be used by other synthesis tools to improve the data path design, as will be described in the next two chapters.

5. Control Step Scheduling (CSTEP)

Scheduling assigns operators in the behavioral specification to control steps that represent clock cycles in the completed design. Because scheduling and other design tasks are heavily interdependent, it has been recognized as an important part of the synthesis problem [Gajski86]. When interface information is added to the synthesis problem, scheduling becomes even more important because it determines whether timing constraints can be met in the resulting implementation.

Scheduling has strong interactions with other parts of the synthesis process because it affects two important factors in the design:

- The *timing* of the design, including both the overall speed of the design *and* the relative timing of operators implemented in the design.

- The *resources* needed in the design i.e. the registers, functional units, and interconnections

This chapter describes CSTEP, a scheduling algorithm that uses techniques drawn from microcode compaction. Unlike other approaches, CSTEP has a primary goal of dealing with interface timing constraints as well as performance and area constraints.

This chapter is divided into several sections. Section 5.1 provides an overview of the scheduling problem and its implications on hardware and timing. Section 5.2 discusses related work in scheduling and contrasts the CSTEP approach to these efforts. Section 5.3 describes the CSTEP approach and presents the modified list scheduling algorithm used by CSTEP including the timing and resource considerations. Section 5.4 gives some examples of using CSTEP's approach to scheduling, and Section 5.5 concludes the chapter.

5.1. THE SCHEDULING PROBLEM

Scheduling assigns each operator x_a to a contiguous range of control steps [cs_{i1}, cs_{i2}], where each control step cs_i corresponds to a clock cycle in the completed design. Scheduling is difficult because of interdependencies. Scheduling has a distinct affect on the hardware needed to implement a design, and in turn is affected by timing constraints which may be imposed by external hardware considerations.

In the simplest case, operators are assigned to a single control step, as shown in Figure 5-1a. Each data operator that is scheduled in a control step implies a combinational functional unit that will implement the operator during the control step. The combinational delay of the functional unit and interconnect must not exceed the clock period. If the combinational delay exceeds the clock period, operators are sometimes scheduled into multiple control steps, as shown in Figure 5-2b. Multiple step operators imply a functional unit (such as the multiplier shown) which implements the operation in multiple control steps.

Data flow dependencies require that an operator that produces a value must be scheduled before an operator that uses that value. When the two operators are scheduled in different control steps, this implies that the result must be stored in a register until it is used. An alternative approach is to schedule the two operators into the same control step, as shown in Figure 5-1c. This implies the use of two functional units which are cascaded and operate in the same control step. In this case, the combined functional unit and interconnect delays

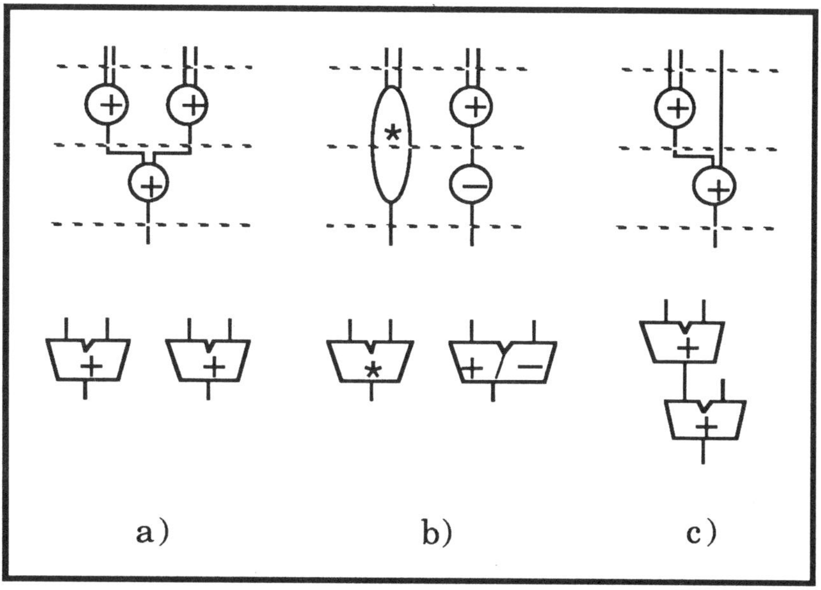

Figure 5-1. Scheduling Alternatives

of both functional units must not exceed the clock period. This form of scheduling is sometimes referred to as *chaining* [Pangrle87].

Scheduling must also deal with control operations such as conditionals, loops, and subroutines. The simplest approach to scheduling conditionals is to divide the flow graph into basic blocks that are scheduled separately using *local scheduling*. For example, Figure 5-2a shows local scheduling for a simple IF-THEN-ELSE construct. The basic blocks corresponding to the THEN and ELSE clauses are scheduled into separate steps. A more complicated approach is to consider multiple basic blocks as a unit--this is referred to as *global scheduling*. A subset of global scheduling, called global slicing [Tseng88], is shown in Figure 5-2b. This alternative schedules operators from both branches of the IF operator into the same set of control steps. These operators are then executed *conditionally* depending on whether the condition of the IF operator is true or false.

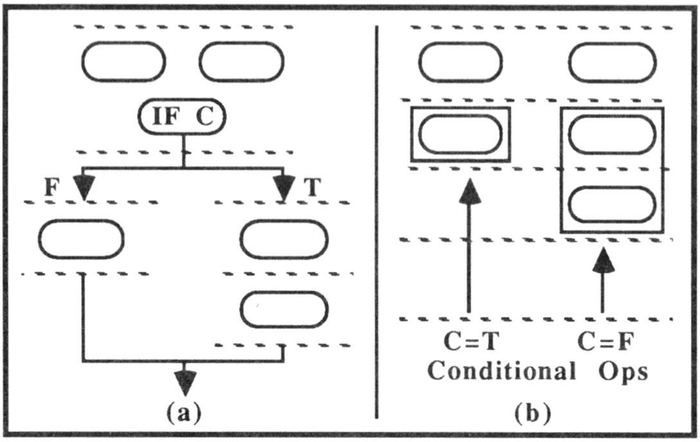

Figure 5-2. Local vs. Global Scheduling of Conditionals

5.1.1. Scheduling and Resources

Scheduling has a major impact on the cost of a synthesized design. While the exact cost cannot be determined until after allocation and binding, scheduling determines lower bounds on the number of registers and functional units required by the design. A lower bound on registers can be established by counting the number of values that require storage in each control step and finding the maximum number over all control steps. A lower bound on functional units can be found by counting the number of similar operators that are scheduled in parallel in each control step and finding the maximum of this value over all control steps. These estimates can be used to guide the scheduling process.

Two useful techniques for exploiting this information are the use of *resource constraints* and *resource utilization*. *Resource constraints* limit the number of resources of each type that are allowed in a design. An operator cannot be added to a control step during scheduling if doing so violates a resource constraint. Resource constraints guarantee that a design will meet a specific register and functional unit cost goal. *Resource utilization* measures the efficiency of hardware in a schedule. A high utilization rate indicates that the control steps are well compacted so the schedule is short relative to the amount of hardware required to implement it. Resource utilization provides a metric of scheduling quality to guide the scheduling process.

5.1.2. Scheduling and Timing

Timing is an important consideration in scheduling for two reasons. First, scheduling fixes the overall timing of a design, which determines both overall performance and the performance of a given set of tasks in the behavioral specification. Second, scheduling fixes the relative timing of individual operators. Relative timing is important when synthesizing circuits that implement predefined *interface protocols* which constrain the relative timing of signaling operators [Nestor87].

Because overall schedule length reflects design performance, most schedulers attempt to minimize overall schedule length. In addition, some approaches allow the use of *performance timing constraints*, which specify timing relationships about the time required to complete a specified set of tasks. In their simplest form, performance constraints specify a maximum length for the overall schedule [Parker86]. More flexible approaches to specifying performance constraints [Girczyc85, Camposano86] allow the specification of minimum and maximum time constraints on arbitrary intervals of statements in the behavioral description. These constraints specify a time for the completion of all of the operators described in the interval.

In contrast, *interface timing constraints* are concerned with the relative timing of a small set of individual signaling operators. Interface protocols specify minimum and maximum values for these operations but do not specify the timing of unrelated operators. For this reason, interface timing constraints are best expressed on individual pairs of operators rather than intervals containing several operators.

CSTEP supports interface timing constraints which specify minimum and maximum times between operations. Constraint declarations in CSTEP describe a constraint name, a time interval between operators, and an inequality with a constant that is expressed in either nanoseconds or clock periods. For example, the following maximum-time constraint specifies that operator x1 must execute at least 80ns before operator x2:

```
c1: time(x1,x2) LSS 80ns;
```

Similarly, the following minimum-time constraint specifies that operator x3 must execute at least one clock period before after operator x4:

c2: time(x3,x4) GEQ 1 clock;

Constraints on equality and inequality are accepted by CSTEP and treated as combinations of minimum and maximum timing constraints.

5.2. RELATED WORK

Numerous approaches to scheduling have been reported in the literature ranging from simple local scheduling heuristics to sophisticated global approaches which try to balance performance against resource usage. This section discusses these approaches and compares them to the CSTEP scheduler.

5.2.1. Local Scheduling

The goal of most early approaches to scheduling was to quickly generate a scheduled data flow graph that could be used by separate allocation and binding tools. As synthesis techniques gained sophistication, more complex approaches were introduced. Most early approaches to scheduling are based on heuristics drawn from research in *microcode compaction* [Davidson81], including *as-soon-as-possible (ASAP)* scheduling and *list scheduling*. In ASAP scheduling, operators are scheduled into increasing control steps as allowed by data dependencies, hardware conflicts, and their initial order. In list scheduling, operators are iteratively scheduled into control steps as allowed by data dependencies and hardware conflicts but are selected for placement in an order that is determined by a heuristic priority function. These techniques are *local* in nature because they schedule only one basic block at a time.

Tseng [Tseng84a] used ASAP scheduling as part of the EMERALD synthesis system. In this system, an initial schedule is produced from the VT before allocation and binding. After allocation and binding, a new schedule is produced, attempting to maximize compaction. Gatenby [Gatenby81] developed a scheduler using a variant of list scheduling that was later improved by Hitchcock [Hitchcock83b]. This scheduler supported both multiple step operators and cascaded operators, and scheduled operations in execution order. This scheduler was used with the initial versions of the EMUCS and DAA synthesis tools [Thomas83].

Other approaches to local scheduling include BUD [McFarland86a] and SLICER [Pangrle87]. BUD performs list scheduling with resource constraints but schedules operations in reverse order. SLICER determines the *critical path* of operations which would increase the length of a schedule if delayed and the mobility of operations not on the critical path (i.e. how much a non-critical operation can be delayed without increasing the length of the schedule). It then schedules operations using list scheduling with mobility as priority function. In this way, operations with the smallest mobility are scheduled first.

5.2.2. Global Scheduling

Because of the simplicity of the heuristics and the small size of most basic blocks, local scheduling techniques are fast and flexible. In contrast, global techniques take more time but provide higher quality schedules. In these approaches, a data-flow graph that includes conditionals is scheduled into a contiguous set of control steps. Recent approaches to global scheduling have concentrated on meeting a performance constraint expressed as a maximum-time constraint on all the operations in the graph while using heuristics that attempt to minimize the hardware resources in the design.

The ELF [Girczyc84] synthesis system uses a global variant of list scheduling called urgency scheduling. Unlike many synthesis systems, ELF also performs allocation and binding at the same time as operators are selected for scheduling. Operators are selected for scheduling and binding using a priority function called an *urgency factor*. The urgency factor reflects the probability of violating the maximum time constraint. A secondary priority function reflects the effect of scheduling the operator on hardware resources, and so attempts to minimize resource usage after satisfying hardware constraints.

MAHA [Parker86] performs operator allocation and binding concurrently with scheduling and schedules operators on the critical path first. It then schedules operators depending on operator mobility (called freedom in this case) and resource constraints.

HAL [Paulin87] uses a heuristic called *force-directed scheduling*. This heuristic schedules operations by considering all possible assignments of operations to control steps and selecting operators based on a "force" value that reflects the potential parallelism of operators and utilization of hardware in the design. By scheduling minimum-force operators, force-directed scheduling reduces the overall parallelism of the design and thus the corresponding resource requirements.

The Yorktown Silicon Compiler(YSC) [Brayton87] takes a very different approach to global scheduling. The YSC schedules as many operations as possible into states, minimizing the total number of states. It then performs a separate *state splitting* task that divides operations in a state into two states to reduce the combinational logic delays in the longest states.

5.2.3. Comparison of CSTEP to Related Works

In contrast to these approaches, CSTEP was developed with the goal of supporting interface timing constraints and resource constraints while remaining fast and flexible. To meet these goals, CSTEP uses list scheduling for simplicity and speed. CSTEP uses a set of priority functions to select operators for scheduling while satisfying both maximum and minimum-time interface timing constraints. This differs from the urgency scheduling approach used in ELF, which only satisfies maximum time constraints that are expressed on intervals of operators. CSTEP may also schedule within the bounds of resource constraints supplied by APARTY or may consider resource utilization in choosing a schedule. The list scheduling approach has a major advantage in simplicity and speed. List scheduling is linear in complexity on the number of operators in a basic block. In contrast, approaches such as force-directed scheduling report a complexity of at best $O(n^2)$ on the *total* number of operators in the behavioral specification, as opposed to the number of operators in a basic block. This difference in speed provides CSTEP with the opportunity to iteratively perform scheduling to explore the design space. Finally, local scheduling also allows full support for extended control constructs such as subroutines. Most global scheduling approaches cannot support these constructs.

5.3. THE CSTEP SCHEDULING APPROACH

CSTEP is a simple heuristic for scheduling with hardware and timing constraints that is based on list scheduling. It uses a priority function to select operators for scheduling that considers both interface timing constraints and resource constraints. Each time an operator is considered for placement into a control step, the timing constraints associated with that operator are evaluated as if they were placed in that control step and also as if they were placed in the following control step. Placement is also evaluated according to resource constraints. Resource constraints may be considered as hard limits, in which case they specify the upper bound on hardware to be used for the design, or soft limits, in which case they specify a lower limit on hardware. Depending on whether any constraints are violated, the priority function returns a value that either allows immediate placement of the operator or forces the placement to be delayed until a later step. The priority function also determines the order of placement, so that the most highly constrained operators are placed first.

CSTEP schedules operators into control steps one basic block at a time. Basic blocks are scheduled in execution order using an execution-order traversal of the control flow graph. This guarantees that when a timing constraint is expressed on two operators that are in separate basic blocks, the first operator in the constraint is scheduled before the second operator is scheduled. This leaves the second operator to be evaluated for placement in terms of how placement affects the constraint. The ordered scheduling of basic blocks also ensures that inter-basic block data dependencies will be satisfied.

5.3.1. The CSTEP Algorithm

Fisher [Fisher81] describes the list scheduling algorithm as it is used in microcode compaction. CSTEP uses list scheduling with a number of modifications that support scheduling for synthesis. First, it supports primitive estimation of combinational delay. Second, it uses this estimation to support multiple cycle operators and cascaded operators. Third, it uses a modified priority function in which negative priority values force placement of operators to be delayed.

Each operation is assigned a delay value. These values are maintained in a database that can be changed to reflect different technologies and previous design iterations. The list scheduling algorithm refers to the operation delay of an operator using the function:

$opdelay(\ x_a\ )$

Delay estimation for cascaded hardware (chaining) is supported by calculating accumulated delay. Accumulated delay is an estimation of the total delay through cascaded operators in a control step. It is calculated recursively: the accumulated delay for an operator x_a that is (or will be) scheduled in control step $cs_{i,l}$ is the operation delay of x_a added to the largest accumulated delay of any operators scheduled in $cs_{i,l}$ that are used by x_a for inputs. The list scheduling algorithm refers to accumulated delay using the function:

$acdelay(\ x_a\ ,\ cs_{i,l}\ )$

In basic list scheduling as is done for microcode compaction, an operator cannot be scheduled in a control step until all of the operators that provide it with inputs are scheduled in previous control steps. An operator that meets these conditions is said to be *data ready*. To support cascaded operator scheduling, this definition must be modified. An operator can be placed in the same control step as an operator that provides it with an input if the accumulated delay does not exceed the clock period. With these modifications, an operator x_a in basic block B_i is data ready for placement into control step $cs_{i,l}$ when for each input $i_{a,b}$ of x_a one of the following conditions is met:

1. $i_{a,b}$ is connected to the output $o_{0,c}$ of an operator x_0 which is scheduled in control step $cs_{i,l}$ AND $acdelay(\ x_a\ ,\ cs_{i,l}\ )$ does not exceed the clock period T_{clock}.

2. $i_{a,b}$ is connected to the output $o_{0,c}$ of an operator x_0 which resides in basic block B_i AND x_a is scheduled in a control step that precedes $cs_{i,l}$.

3. $i_{a,b}$ is connected to the output $o_{0,c}$ of an operator x_0 which resides in a basic block other than B_i.

4. $i_{a,b}$ is connected to a vtbody input $o_{0,c}$. Vtbody inputs represent the flow of values such as procedure parameters that must be stored across call operators.

5. $i_{a,b}$ is connected to a constant.

Another important consideration in scheduling is determining when hardware conflicts prevent operators from being scheduled into a control step. If a control step contains operators that are bound to certain hardware elements, additional operators can be scheduled into the same control step only if they do not use the same hardware. More formally, an operator x_a is *resource compatible* with control step $cs_{i,l}$ if for each operator x_0 already scheduled in $cs_{i,l}$ both of the following conditions are met:

1. x_a and x_0 are not yet bound to functional units, or are bound to different functional units.

2. For each output $o_{a,c}$ of x_a and each output $o_{0,c0}$ of x_0 , $o_{a,c}$ and $o_{0,c0}$ are not yet bound to registers or are bound to different registers.

These conditions can be extended to consider operators that are not actually bound to hardware, but that might be implemented by the same hardware unit. CSTEP does not operate on pre-bound hardware, but it can determine whether operators are candidates to share hardware and treat them appropriately. CSTEP does not consider register sharing explicitly.

In order to consider resource constraints during scheduling, assumptions about which operations can be merged into a shared functional unit must be made. The EMUCS data path allocator makes these decisions according to the cost of combining different operations. These costs are listed in the Functional Cost table which is described fully in Section 6.2.1. The Functional Cost table is a Workbench input that reflects the target technology for the design. Costs for combining each pair of operations into a single functional unit are listed in this table. CSTEP uses this table by considering the cost of combining operations relative to a "combine threshold". If the cost of combining two operations is less than the threshold, the scheduler can assume that they may both be implemented by the same hardware unit and should be treated accordingly.

List scheduling is guided by a priority function that determines the order in which data ready operators are selected for placement. In basic list scheduling, this function returns a positive integer that is based on an arbitrary heuristic such as the height of the operator in the data flow graph. The CSTEP scheduling algorithm uses a modification to this approach in which the priority function may delay operator placement by returning a negative value.

To address both timing constraints and a range of resource constraints, the CSTEP priority function is broken down into three separate functions: the primary, secondary, and tertiary priority functions. The primary priority function evaluates timing constraints for an operator and attempts to satisfy timing constraints by either forcing or delaying scheduling. The secondary priority function evaluates resource constraints and delays placement of operators which violate these constraints. The tertiary priority function attempts to maximize the resource utilization in the absence of hard timing and resource constraints.

Figure 5-3 shows the modified list scheduling algorithm. This algorithm uses a counter variable *step* to reference the current control step, three operator lists *oplist*, *drlist*, and *mulist* to keep track of the operators being scheduled, and two hardware lists *maxlist* and *usedlist* to keep track of hardware usage. The *oplist* variable initially contains all the operators in a basic block B_i. As scheduling proceeds, operators that become data ready are removed from *oplist* and placed in the data ready list *drlist*. As operators are selected for scheduling, they are removed from *drlist* and placed in control steps. Multiple step scheduling is supported by the *mulist* and is described below.

Maxlist contains the maximum number of hardware units allowed for implementing any control step. The values in *maxlist* are user inputs that may be modified during scheduling, as described in the next section. *Usedlist* keeps track of the number of hardware units required to implement the current control step. As operators are scheduled, the hardware required to implement them is marked in this list. The number of hardware units listed in *usedlist* for any control step must be less than or equal to the number of hardware units listed in *maxlist*. Each operation is categorized according to the other operations with which it can share hardware, as determined by the Functional Cost table. The categorization defines an operation's *OPTYPE*. Both *usedlist* and *maxlist* are indexed by *OPTYPE*.

```
step = 0;
while (oplist, drlist, and mulist are not empty) {
    clear usedlist
    step = step + 1; create new control step cs_{i,step}
    if ( mulist not empty ) {
        for each operator x_am in mulist {
            schedule x_am into cs_{i,step};
            update usedlist
            if ( x_am is scheduled for longer than opdelay( x_am )){
                    remove x_am from mulist;
                    }
            }
        }
    do{
        Search oplist for data ready operators, remove
                    them from oplist, and place them in drlist;
        Apply the priority function to each operator in drlist and
                    find the operator x_a with the largest priority.
        if ( priority(x_a) ≥ 0 ) {
            remove x_a from drlist;
            schedule x_a into cs_{i,step};
            if ( opdelay(x_a) > T_clock ) {
                    add x_a to mulist;
                    update usedlist
                    }
            }
    } while ( drlist not empty && priority(x_a) ≥ 0 );
}
```

Figure 5-3. The CSTEP List Scheduling Algorithm

Scheduling proceeds iteratively until all three operator lists are empty. The outer loop of the algorithm creates new control steps in increasing order and continues the scheduling of multiple step operators that were started in preceding steps. The inner loop of the algorithm selects operators from the data ready list *drlist* and schedules them into each control step until no further operators can be placed in that control step. As operators are scheduled, new operators become data ready and are removed from *oplist* and placed in *drlist*.

Multiple step scheduling is supported using the *mulist* variable. When an operator is chosen for scheduling, it is tested to see if its delay is longer than the clock period. If so, it is added to *mulist*, which keeps track of operators scheduled in multiple control steps. Each time a new control step is created, operators in *mulist* are scheduled first. Each operator remains in *mulist* until the number of control steps in which it is scheduled exceeds the operator delay.

The extended definition of the data ready condition supports the scheduling of cascaded operators by allowing cascaded operators to be scheduled if they have an accumulated delay that is less than the clock period.

5.3.2. Timing Constraints: the Primary Priority Function

The CSTEP priority functions determine the order in which operators are scheduled into control steps by evaluating the effect of scheduling an operator on timing and resource constraints. Depending on this evaluation, the priority function returns a value that reflects the importance of scheduling this operator with respect to other data ready operators. By returning a large positive value, the priority can force the immediate scheduling of an operator. By returning a negative value, the priority function can delay placement. Operators which are not affected by the primary priority function are passed to the secondary priority function, and operators which are not affected by the secondary priority function are passed in turn to the tertiary priority function.

Timing constraint evaluation is used as the primary priority function. For an operator x_a that is placed in a control step, timing constraints which involve x_a relative to previously scheduled operators are evaluated by counting the number of control steps between them. This value is compared to the value of each constraint to determine whether the constraint is satisfied.

During the execution of the CSTEP algorithm, the primary priority function evaluates data-ready operators for scheduling into the current control step $cs_{i,l}$. It first considers the impact of this placement on minimum-time constraints. If scheduling an operator into the current control step violates any minimum-time constraints, this placement is considered *too early* and must be delayed to a later control step. This is accomplished by returning a negative priority value -BIGINT which delays placement of the operation.

Next, the primary priority function considers the impact of *not* placing this operation immediately on maximum-time constraints by evaluating maximum-time constraints as if x_a were placed in an immediately following control step $cs_{i,l+1}$. Violation of a constraint means that any delay in scheduling this operator would violate the maximum-time constraint. In this case, the priority function forces immediate scheduling by returning a large positive value +BIGINT. This value is larger than any values returned by other priority functions and forces the selection of this operator for scheduling.

In the case where it is acceptable to schedule an operator, but not imperative to do so immediately, the decision will be passed on to the secondary priority function.

Timing constraints and control flow. Timing constraints may be placed on any pair of operators in a behavioral specification. Evaluation of timing constraints must consider the possibility that two operations may be separated by conditionals and loops. In the case of conditionals, there may be more than one path of execution between the two operators. For example, Figure 5-4 shows two operations separated by a single IF-THEN-ELSE conditional construct. Since the two parts of the conditional may take different amounts of time to execute, any timing constraint on the two operators must be satisfied when either of the two paths is executed. When conditionals are nested, additional paths of execution are possible. CSTEP deals with this possibility in a preprocessing step that enumerates the paths between operators and creates a separate constraint for each path.

Dealing with loops is a more complicated problem, because in general the number of iterations that a loop executes is not known. Currently, CSTEP evaluates the worst case for minimum-time constraints by assuming that the loop executes a minimum of one iteration. If lower and upper bounds can be supplied by the user, it is straightforward to evaluate both minimum and maximum-time constraints using this worst-case information.

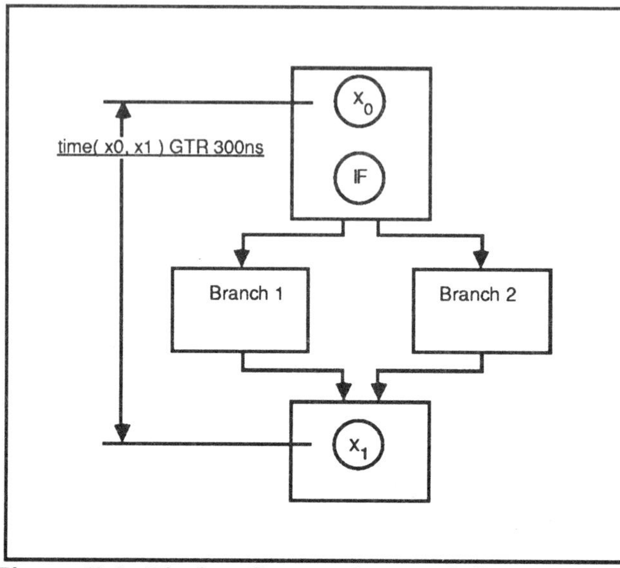

Figure 5-4. Timing Constraints and Control Flow

Constraint Interactions. List scheduling operates by placing operators into successive control steps. Data-flow constraints, control-flow constraints, and minimum-time constraints can all have the effect of delaying the scheduling of an operator until these constraints are satisfied. If the delayed operator is also subject to a maximum-time constraint, then a possible conflict exists between the two constraints that may not be resolved by the CSTEP algorithm alone.

For example, Figure 5-5 illustrates a possible conflict between a data-flow constraint and a maximum-time constraint. In this case, there is a data flow constraint between operators x_1 and x_2. Because of this constraint, x_2 cannot be scheduled until x_1 is scheduled. However, x_2 may also be subject to a maximum-time constraint on a third operator, x_0. The data flow constraint can delay placement of x_1 without consideration of the maximum-time constraint because x_1 is not involved in the maximum time constraint. This lack of consideration may cause the maximum-time constraint to be violated.

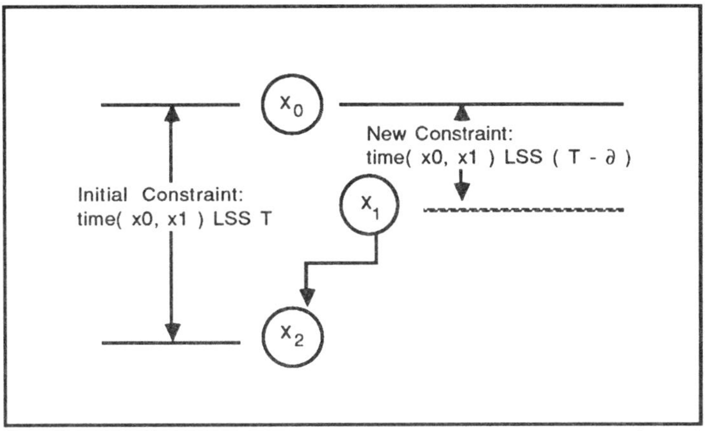

Figure 5-5. Constraint Interaction for Data Dependencies

To deal with this eventuality, CSTEP examines the data-flow constraints on the second operator in each maximum-time constraint prior to scheduling. The potential conflict caused by a data flow constraint and a maximum-time constraint is remedied by placing a new maximum-time constraint on each operator that produces a value used by the original constrained operator. The value of this maximum-time constraint is taken from the initial maximum-time constraint and is adjusted by a value ∂ that guarantees that this operator will be scheduled in time to prevent the original maximum-time constraint from being violated. Typically, ∂ is chosen to be the combinational delay of the original constrained operator. In the example of Figure 5-5, a new constraint is created between $x0$ and $x1$ that guarantees that the constraint between x_0 and x_2 is not violated.

In the case of control-flow constraints, conflicts can arise when operators in a maximum-time constraint reside in separate basic blocks, as shown in Figure 5-6. In a local scheduling algorithm like CSTEP, operators in a basic block are not scheduled until operators in all preceding basic blocks are scheduled. In the example of Figure 5-6, all the operators in basic block B_2 can delay the placement of operators in later basic blocks, including the constrained operator x_3. Since these operators are not constrained in a maximum-time constraint, they can delay the scheduling of x_3 until the constraint is violated. This can be avoided by creating new constraints on each operator in the separating basic blocks.

In the final case of minimum-time constraints, conflicts can arise when the constrained operator also has a maximum-time constraint, as shown in Figure 5-7. In this case, constraint C1 requires that the relative time between operators x_1 and x_2 be greater than than time T1. At the same time, constraint C2 requires that the relative time between operators x_0 and x_2 be less than time T2. As in the other cases, a new constraint must be added to prevent a conflict between the two constraints. However, the value of the new constraint is now the difference between the maximum-time constraint and the minimum-time constraint. If this difference is positive, then the new constraint must be treated as a maximum-time constraint. However, if it is negative then the new constraint must be treated as a minimum-time constraint. This guarantees that both of the original constraints will be satisfied in scheduling.

Creating new constraints due to constraint interactions is a recursive process, in which new constraints may themselves be subject to constraint interactions. Because of this, a large number of new constraints may be created. However, this concern is mitigated by the local nature of most timing constraints. Even if the constraints are not local, only the strictest constraints on each pair of operators needs to be maintained - other less strict constraints can be discarded, reducing the total number of constraints.

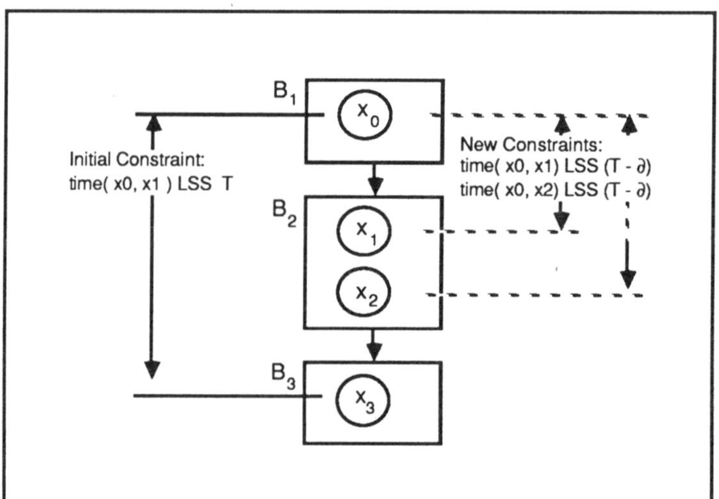

Figure 5-6. Constraint Interaction for Control Dependencies

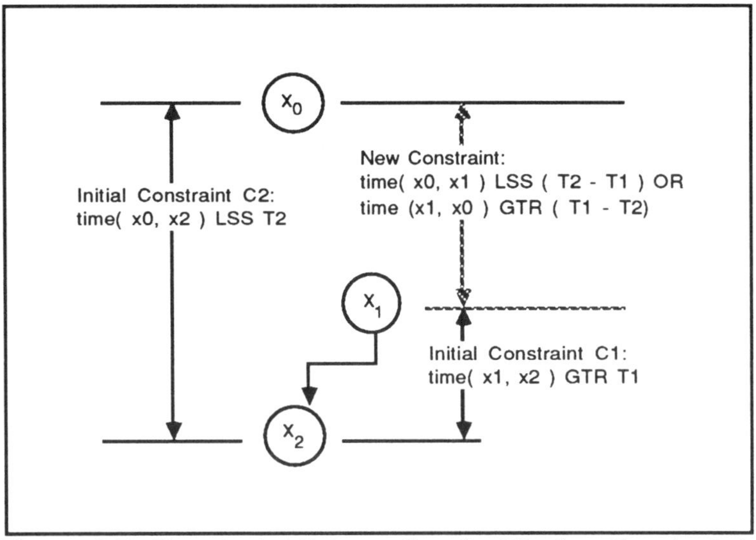

Figure 5-7. Constraint Interaction for Timing Constraints

5.3.3. Resource Constraints: The Secondary Priority Function.

If timing constraints have been satisfied and there is still some question as to whether to schedule an operator, the secondary priority function is invoked to consider resource constraints. As scheduling proceeds for a particular control step, and operators are assigned to that control step, the need for a particular type of functional unit (implementing a particular operator of type *OPTYPE*) is marked in the *usedlist*. (i.e. When an operator of type *OPTYPE* is scheduled in the current control step, *usedlist[OPTYPE]* is incremented.) During the scheduling process, when considering a data ready operator x which executes operation i which is also of type *OPTYPE* for placement in the control step, *usedlist* is compared with *maxlist*. If *usedlist[OPTYPE]* (the number of functional units required to implement operators of type *OPTYPE* in this control step), is less than to *maxlist[OPTYPE]* (the maximum number of functional units implementing operators of type *OPTYPE*), then operator x may be scheduled in the current time step and a positive value is returned by the priority function. This value is determined according to the importance of placing the operator early in the scheduling process.

The later a data ready operator is considered for scheduling in a given control step, the greater the possibility that it will be delayed due to resource constraints because other operators of the same type may already have been assigned to the given control step, thereby using all of the available functional units. Therefore it is vital to the creation of a good schedule to consider the most important operators first. Determining those important operators is a difficult problem. It is generally best to schedule those operators that have many other operators dependent on their outputs early, although it is often difficult to foresee when the scheduling of an operator may block a better move several control steps later. CSTEP uses 2 heuristics to choose which data ready operators to consider first for scheduling.

The first of these heuristics looks at the depth of the chain of data dependent operators linked to the operators under consideration. This heuristic essentially looks for the critical path and attempts to bind operators in that path first. The priority value returned by this heuristic is the dependency depth.

In the case of two operators having the same length data dependency chain, the second heuristic is considered. The second heuristic counts the total number of operators that are data dependent on the operators under consideration. An operator with a larger dependency tree is more likely to cause a bottleneck, so gets a higher priority.

If the secondary function determines that placement of the current operator would violate the maximum hardware specifications, one of two actions will take place. If the user has chosen to run in "hard mode," the maxlist values are accepted as hard limits that cannot be violated by the scheduler, so a negative value is returned by the priority function so that placement is delayed. If, however, the user has chosen "soft mode," wherein the initial maxlist values imply a lower limit on hardware and CSTEP may choose higher limits according to utilization, the tertiary priority function is invoked.

5.3.4. Resource Utilization: The Tertiary Priority Function.

While scheduling does not actually allocate hardware, it does determine the minimum number of functional units and registers required to implement the data path. One consideration in determining how much hardware is appropriate for a particular design is utilization. It is advantageous to maximize the efficiency of each piece of hardware by making sure that each piece of allocated hardware is used as much as possible. Soft mode considers the utilization of functional units in determining the schedule.

Utilization should be measured across the whole schedule. However, since CSTEP does not know the length of the schedule during the scheduling process, it cannot get an exact measure of utilization for any functional unit. It can consider the utilization of a functional unit over a small window of the schedule, and it uses this window to make decisions about how much hardware to use.

In soft mode, the maxalloc list is not fixed, but its function remains the same: it lists the amount of hardware available for use in scheduling. The utilization checker considers all of the operators that would be data ready in the next control step if an additional (*maxlist[OPTYPE] + 1*) piece of "hardware" were allowed in the current control step. If the data ready operators for the next control step can utilize *maxlist[OPTYPE] + 1* pieces of "hardware" (i.e. this piece of hardware will be used in greater than 1 control step), then *maxlist[OPTYPE]* is incremented, and a positive value is returned for the operator in question. If new hardware cannot be utilized in at least two consecutive control steps, a negative value is returned. In either case, scheduling continues from this point.

Figure 5-8 shows how this resource utilization checker works in the scheduling process. In the figure, the PLUS operators on the right are data ready for the control step indicated in **bold** print at the left and have been ordered according to their importance. In the first control step, three operators are data ready. Assume that the initial maxlist value for the PLUS operation is 1, so that at least one plus operator can be scheduled in each control step. The figure shows that operator $+_a$ has been scheduled.

The next row of the figure shows that $+_b$ is the next operator to be considered for scheduling. Since the maximum number of prespecified functional units have already been used in the first control step, the resource utilization checker must be invoked. Operator $+_b$ is temporarily placed in control step 1, and then the next control step is examined. Control step 2 shows that if $+_b$ were to be placed in control step 1, $+_c$ which was data ready for control step 1 could be placed in control step 2, and $+_d$, which would be made data ready for control step 2

($+_d$ is data dependent on $+_b$) could also be scheduled. Since a second functional unit could be utilized in both control steps, the *maxlist* value for the PLUS operation is incremented and $+_b$ is placed in control step 1.

Now $+_c$ is considered for placement in control step 1. As shown in the third row of the figure, if $+_c$ is placed in control step 1, there are only 2 PLUS operators that can be placed in control step 2. Since a third functional unit cannot be utilized in both control steps 1 and 2, the placement of $+_c$ is delayed, even though it is data ready. Since there are no more operators to be considered for placement in control step 1, scheduling proceeds to control step 2. At this point both operators $+_c$ and $+_d$ are data ready, and because the maxlist value for PLUS is 2, both can be scheduled without further consideration for resource utilization. Operator $+_e$, which is made data ready by operator $+_c$ must be delayed until control step 3. The final schedule is listed in row 4 of the figure.

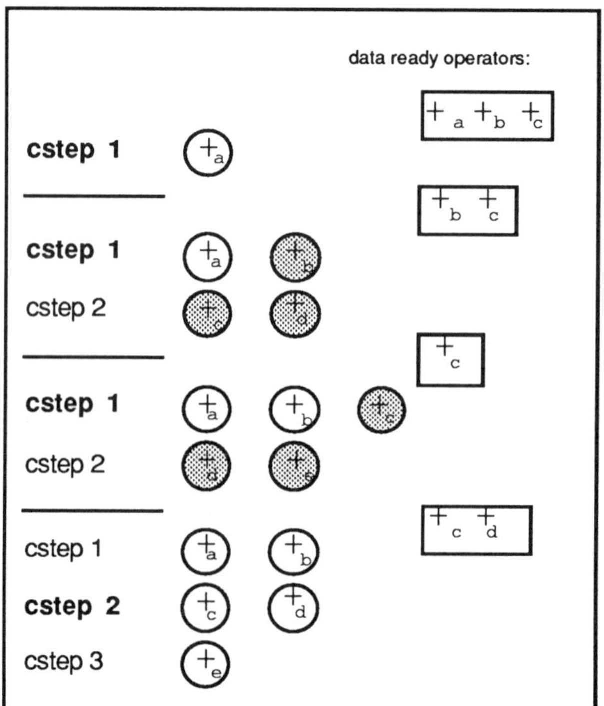

Figure 5-8. Using Resource Utilization to Determine Hardware Limits

Iteration for a Better Schedule. As scheduling progresses, *maxlist* may increase for some or all OPTYPES, until it reaches the maximum that CSTEP will allow it to be for the design. At this point, the control steps scheduled are as parallel as CSTEP will allow them to be. This state will be referred to as "As Parallel As Possible," or APAP. (Note that APAP may not be maximally parallel depending on the hardware limitations set by the user.) Control steps scheduled before maxlist has reached its maxima may not be APAP. Therefore, re-running CSTEP with *maxlist* initialized to its maxima will allow the early steps to be scheduled APAP relative to the "hardware" that is allocated for the schedule. This re-scheduling may reduce the total number of control steps required for the schedule. (It is possible that the schedule would remain the same length.) By defining *maxlist* as a hard limit in the second pass, the number of "hardware units" cannot increase. Thus the schedule cannot get worse with the second pass.

Partitioning Information and Resource Considerations. It is easy to take partitioning information into account when building the schedule. The information from the partitioner is passed on to CSTEP through the VT. VT allows additional information about its components to be included in the form of qualifiers. The partitioner qualifies each operator with a number indicating the partition in which it belongs. Thus the VT that CSTEP uses contains all of the partitioning information that it needs.

To take this partitioning information into consideration, CSTEP requires a separate *maxlist* array for each partition. Then the determination of whether an operator can be scheduled proceeds as described above, but the decision is made relative to the *maxlist* array that is associated with the partition that contains the operator under consideration.

5.4. SCHEDULING EXAMPLES

Two scheduling examples are given below. Timing and resource constraints are depicted in separate examples for clarity.

5.4.1. A Timing Constraint Scheduling Example

To illustrate the action of scheduling with timing, we now present an example from the ISYN system [Nestor87], a set of synthesis tools built on top of the Workbench for synthesis of circuits with complicated interface requirements. The following ISPS fragment and timing constraints describe the behavior of a simplified version of the Intel Multibus 1 [Intel79] performing a bus write operation:

```
OUTPUTP( adr.l, TS.ENABLE )  {L:mw1e};
OUTPUTP( data.l, TS.ENABLE ) {L:mw2e};
OUTPUTP(adr.l, Address)      {L:mw1} ;
OUTPUTP(data.l, Data)        {L:mw2} ;
OUTPUTP(mwtc.l, 1)           {L:mw3} NEXT
WAIT(INPUTP(xack.l)) ;
OUTPUTP(mwtc.l, 0)            {L:mw4} ;
OUTPUTP(adr.l, TS.DISABLE)   {L:mw5} ;
OUTPUTP(data.l, TS.DISABLE)  {L:mw6} ;
```

```
Tena:         time( mw1e, mw1 ) EQL 0 ns;
Tenb:         time( mw2e, mw2 ) EQL 0 ns;
Tas:          time( mw1, mw3 ) GEQ 50ns;
Tds:          time( mw2, mw3 ) GEQ 50ns;
Tdh:          time( mw4, mw5 ) GEQ 50ns;
Tah:          time( mw4, mw6 ) GEQ 50ns;
Tcmd:         time( mw3, mw4 ) GEQ 100ns;
```

INPUTP and OUTPUTP operators specify when values are read from and written to the external ports of the bus interface. OUTPUTP operators can be used with symbolic constants to enable and disable tristate output ports. The timing constraints in the specification are expressed on operators by adding labels using ISPS qualifiers and referring to these labels in the timing constraint specification.

When this specification is translated to a VT and scheduled with CSTEP, the resulting schedule appears as shown in Figure 5-9. This schedule meets all the required timing constraints through its placement of operators into control steps. Timing constraints specify that the first four operators that enable and assert values on the ADR.L and DATA.L ports must execute at least 50 nanoseconds before the command signal MWTC.L is asserted. This causes CSTEP to delay placement of this operator to the second control step. The third control step contains a wait loop which tests the value of XACK.L. The fourth control step deasserts MWTC.L. Additional timing constraints require that the address and data signals must be disabled at least 50 nanoseconds after MWTC.L is deasserted, forcing the final two operators to be scheduled in the fifth control step. More details on this and other interface examples can be found in [Nestor87].

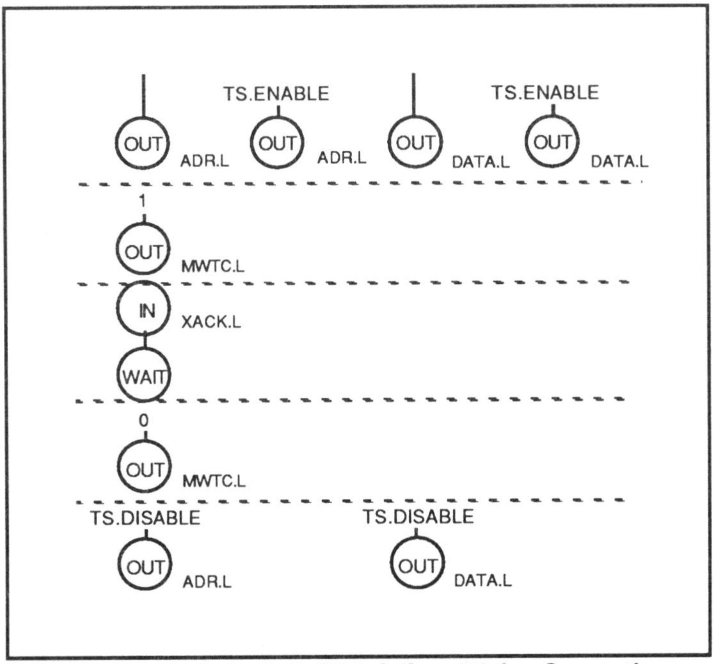

Figure 5-9. Scheduled Multibus Write Operation

5.4.2. A Resource Constraint Scheduling Example

The Elliptical Filter presented in Chapter 3 and for which partitioning results are presented in Chapter 4 is an interesting candidate for resource constraint information because there are many possible schedules for the data-flow. Table 5-1 shows the results of using various resource options to make schedules for the filter example.

The table shows scheduling results from both HAL [Paulin87] and CSTEP. The major difference between HAL's approach to the scheduling problem and CSTEP's approach is that while HAL normally works from a fixed schedule length, CSTEP works with an open-ended schedule, determining the length by the schedule it chooses. This difference is represented in the table by the **bold** numbers which indicate the constraints on the schedulers. The results produced by HAL are presented in the first 5 columns of data. Given a schedule length, HAL chooses a schedule which determines the minimum number of adders and multipliers required to implement the data path, as indicated in the table. All schedules are optimal except for the 19 control step schedule which could be done with one less control step while using the same number of adders and multipliers.

The sixth column of data shows the results of using CSTEP without any resource constraints to produce a schedule. The schedule is of minimum length, but requires more than the minimum hardware for a 17 control step schedule. Soft mode, as shown in column 7, suggests that 3 adders and 3 multipliers be used for the design. Soft mode does not produce a minimum length schedule for its chosen levels of hardware, but using the suggested hardware constraints in hard mode does produce an optimal length schedule. Several hard mode results are listed in columns 8-12. The schedule lengths match those produced by HAL for the given hardware limits.

As the scheduler for the Workbench, it is appropriate for CSTEP to use a fixed hardware approach to scheduling rather than a fixed schedule length because it can use the high level structural information provided by the partitioner to help determine an appropriate number of functional units. Chapter 8 describes the synthesis of the Elliptical Filter. The partitioner suggests a 2 partition design which implies a minimum of 2 adders and 2 multipliers, producing a 19 control step schedule, as listed in Table 5-1. This schedule will be used by EMUCS and is used in Chapter 8 to determine a data path for the elliptical filter.

Table 5-1
Using Resource Constraints to Guide the Scheduling of the Elliptical Filter

	HAL					CSTEP						
						no rsrc	soft	hard				
Control Steps	**17**	**18**	**19**	**20**	**21**	17	21	19	18	18	17	21
Plus	3	3	2	2	2	4	3	**2**	**4**	**3**	**3**	**2**
Multiply	3	2	2	2	1	4	3	**2**	**2**	**2**	**3**	**1**

5.5. SUMMARY

This chapter has described the design and implementation of CSTEP, the Workbench scheduling tool. CSTEP was designed for speed and flexibility and supports scheduling with resource constraints and interface timing constraints.

6. Data Path Allocation (EMUCS)

Data path allocation encompasses several steps of the design process: the allocation of hardware, the binding of data flow operators and values to that hardware, and the appropriate interconnection of the hardware to realize the data-flow. The hardware allocated is at the functional block level and includes modules such as registers, functional units, and multiplexors. In general, Value Trace operators are bound to functional units, VT values are bound to registers or wires, and multiplexors and busses are used to steer data flow among the other components. The goal of data path allocation is to build a functional block level design that is both small and easily realizable with respect to the target technology.

This chapter describes how the EMUCS program can be used to choose a functional block implementation of a data path given a scheduled data-flow graph and information characterizing the target technology. The original EMUCS algorithm is due to McFarland [Nestor87], and was implemented by Hitchcock [Hitchcock83a, Hitchcock83b]. Both the algorithm and the implementation have since been modified to be more general. The current implementation of the EMUCS algorithm is described in this chapter. A more formal presentation of the translation from behavior to structure was presented in Chapter 2.

The data path design is constrained by the control step schedule that it implements. All operators in the schedule must be bound to data path hardware, but operators that are active simultaneously cannot share hardware. Values that are alive across control step boundaries must be stored in registers. These values are called *regvals*, and must be bound explicitly. Values that do not require storage are implemented as the wires that interconnect the operators that produce them with the appropriate destination modules. These values do not need to be bound explicitly. Figure 6-1 shows a schedule and the data flow for that schedule. In the figure, the value C does not require storage because it is not used outside the cstep in which it is created. The value D, however, will require storage because it is active across the boundary of csteps 1 and 2. Therefore D is a regval that will be bound to a register by EMUCS. As is the case for operators, regvals that are active simultaneously cannot share hardware.

cstep #	operations
cstep 1	C = A + B; D = C + 1;
cstep 2	E = D + A;

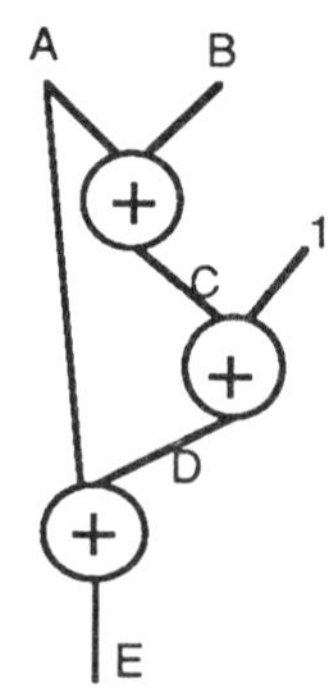

Figure 6-1. Values Alive Across Csteps Require Storage

A small control step schedule and two data path implementations for that schedule are shown in Figure 6-2*. The schedule has 2 PLUS operators that may be implemented by a single adder since they are not active simultaneously. This possible implementation is shown in part *a* of the figure. The implementation requires a multiplexor at each adder input in order to choose the appropriate input for each PLUS

---

* Refer to Figure 2-7 for a key to the symbols used in the data path drawings in this chapter.

operator. The implementation shown in part *b* of Figure 6-2 eliminates the need for multiplexors by using different adders for each add operation and assumes that the final C value will be stored elsewhere if necessary. The quality of the two designs must be evaluated in terms of the target technology. In general, the "best" of these two designs can only be determined with a knowledge of the relative costs to realize multiplexors and functional units.

cstep #	operations
cstep 0	A = 2; B = 3;
cstep 1	C = A + B;
cstep 2	C = C + A;

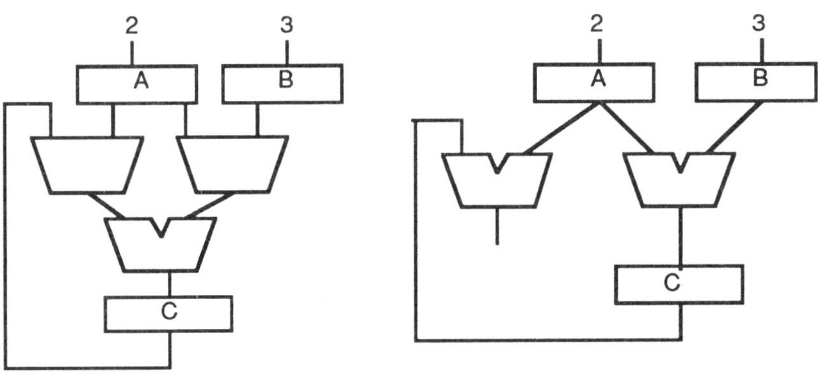

a) Single adder implemetation b) Double adder implementation

Figure 6-2. Two Data Path Implementations of a Single Schedule

6.1. OTHER DATA PATH ALLOCATORS

Before discussing the specifics of the EMUCS data path allocator, several other data path allocators will be described briefly. McFarland [McFarland88] provides a tutorial on High-Level Synthesis in which he defines two classes of data path allocators: iterative/constructive and global allocation. Iterative/constructive techniques bind one element at a time, while global allocation techniques find simultaneous solutions to a number of bindings at one time. Examples of each of these techniques are described below.

6.1.1. HAL

HAL [Paulin86] is a synthesis program made up of several modules that implement synthesis subtasks. The MidHAL module uses a rule-based expert system to choose functional units for a scheduled data-flow. After functional units have been chosen from a cell library according to costs and timing constraints, the ExHAL module creates the data path. Using an iterative/constructive technique, the ExHAL module assigns operators to functional units, allocates and binds registers, and creates the appropriate interconnect. ExHAL uses optimizing steps to improve register utilization and align operands to reduce multiplexor inputs.

6.1.2. MIMOLA

MIMOLA [Marwedel86] has an algorithmic approach to synthesis. After a schedule has been created, a data path is designed in three algorithmic sub-phases: register assignment, module selection, and interconnect generation. Module selection is done with a global allocation technique, while the other two phases use iterative techniques.

6.1.3. CAMAD

The CAMAD system [Peng86] views synthesis as a series of optimizations. Optimizations are iteratively applied to the data-flow to do such things as merging operators together so that they share hardware, or serializing parallel operators so that they can be merged. The choice of which optimizations to apply and the order in which to apply them is determined by hardware costs.

6.1.4. ADAM

In the ADAM system, functional units are chosen by MAHA [Parker86]. In this system, functional units are allocated concurrently with the creation of the schedule. Functional unit allocation falls into the category of iterative techniques. Functional units are allocated in a first-come-first-served basis, with operators sharing functional units when possible. MAHA allocates additional functional units as they become necessary to the implementation of the schedule. Registers are chosen later using a global allocation technique implemented by REAL [Kurdahi87], another component of the ADAM system.

6.1.5. Chippe

Chippe [Brewer87] provides a design expert which chooses functional units from a data base and selects a design strategy to be followed by its synthesis tools, Slicer and Splicer. Slicer chooses a schedule for the design. The Splicer program binds operators to the chosen functional units and creates the appropriate interconnect according to constraints imposed by the expert. The depth first search used by Splicer allows it to find an initial solution quickly so that the expert can direct iterative refinement of the design.

6.1.6. DAA

DAA [Kowalski84] is a rule-based expert system approach to synthesis. The choice of functional units, however, is determined algorithmically by BUD [McFarland86a], which estimates area and performance for several design options, and chooses the best. The DAA program chooses registers and an appropriate interconnection scheme for the design.

6.1.7. Emerald

Emerald [Tseng86] uses a clique partitioning algorithm to choose globally the various segments of the designs such as registers, functional units, and interconnect. The clique partitioning algorithm is described in Section 6.6.1.

6.2. EMUCS OVERVIEW

EMUCS is an algorithmic data path allocator that uses an iterative/constructive technique and can be tuned to different target technologies. The tuning is accomplished through the use of input cost tables that reflect the costs of realizing certain design tradeoffs. The costs are used to make design decisions in several of the five design phases that make up the EMUCS program. Those phases are: initialization of data structures, prebinding of special operators and regvals, automatic binding of all other operators and regvals, a post-processing phase to improve the design, and a finish-up phase to clean up the final design. Of these phases, the automatic binding phase is the most complex. It uses a one-step-lookahead algorithm that considers relative costs for making particular bindings to choose the order in which the VT operators and regvals are bound. EMUCS binds the chosen operator or regval to the module that will result in the lowest cost increase for the design. Before describing the data path allocation phases, some preliminary information is provided.

6.2.1. Technology Input Files

Information about the target technology for the design is provided to EMUCS through two input cost files. One cost file, the Interconnect/Register Cost table contains relative costs of implementing interconnect and registers. The other file, the Functional Cost table, contains relative costs for implementing functional units and for implementing multiple functions within a single functional unit.

Interconnect/Register Cost Table. The Interconnect/Register Cost table contains 6 costs that can be varied by the user to reflect the relative costs of interconnect and registers in the target technology. These cost values are listed below.

- **direct-connection** — the cost of making a direct connection (wire) between any two modules
- **add-input-to-multiplexor** — the cost of adding a new input to a multiplexor
- **change-input-to-multiplexor** — the cost of reconnecting a direct connection through a multiplexor (the cost of adding a multiplexor to the design)
- **add-output-connection** — the cost of adding a new output connection to a module
- **new-register** — the cost of allocating a new register

Functional Cost Table. The Functional Cost table allows the user to specify the similarity of operators according to the particular technology or cell library being used. The Functional Cost table is a two dimensional matrix of values, each representing the cost of adding new functionality to a functional unit. The cost of adding operation n to a functional unit that can already execute a function m is listed in the table as the value in column n, row m.

An extra row at the end of the table is used to represent an empty or unallocated functional unit. The values in this row represent the costs of allocating new functional units for each operation. Thus the cost of allocating a new functional unit with functionality *n* is contained in the *nth* column of the last row of the table. If a functional unit implements multiple functions, the cost of adding new functionality to that functional unit is assumed to be the minimum of adding the new functionality to any one of the functions already implemented by the functional unit.

6.2.2. How Binding Is Accomplished

Two of the EMUCS design phases involve binding data-flow objects to data path modules. Before considering the specifics of the prebinding and automatic binding phases, it is important to understand how the binding is accomplished.

Bindings may be chosen automatically or manually, but the binding itself is always done automatically. Several things must be accomplished by the binder. Links must be established between the data-flow object being bound and the module to which it is being bound. This is demonstrated in Figure 6-3. Interconnections must be made with any appropriate modules, and modules and input ports must be tagged with activation times that will be extracted to generate the control sequence file for the data path.

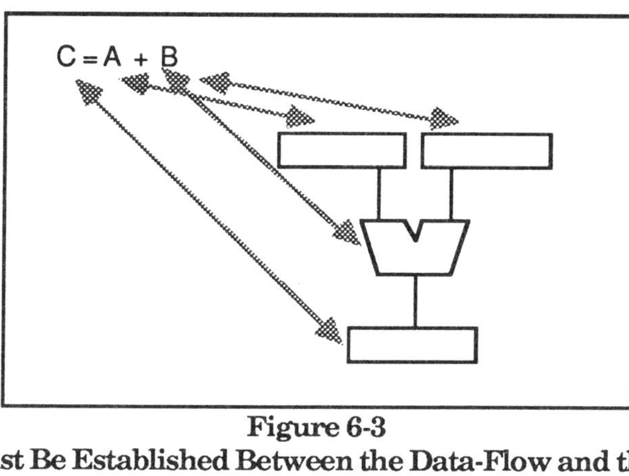

Figure 6-3
Pointers Must Be Established Between the Data-Flow and the Data Path

Sections 6-3 through 6-8 describe the various phases used by EMUCS to design a data path.

6.3. INITIALIZATION

During initialization, EMUCS gathers the information that it needs in order to produce a data path. The user may choose to have EMUCS operate on a complete description, or on individual processes, as determined by behavioral transformations (see Chapter 3). The initialization phase gathers only the information needed for the process specified by the user.

In the initialization phase, EMUCS uses the schedule produced by CSTEP to find all of the operators and all of the regvals that need to be bound and determines their lifetimes through lifetime analysis. The regvals are all those values that require storage. The VT operator and value structures are annotated to allow for the storage of information relevant to EMUCS. Some of this information is later used by CORAL to determine relationships between design representations, as described in Chapter 9. This storage includes space to mark which module the operator or regval gets bound to as well as its partition number.

The initialization phase is also used to make a table of conflicts between the regvals. Each pair of regvals that cannot be stored in the same register due to timing conflicts is marked and the information is stored in a hash table for easy access during processing.

Also during initialization, the interaction of operators and regvals in the VT is examined. A "connection" structure is associated with each operator and regval, listing the other operators and regvals that must be connected to the inputs and outputs of the data flow item. This is done to improve the speed of the algorithm by eliminating the need to search the VT each time a cost calculation or binding is made. This has the added benefit of maintaining consistency by concentrating the search of the VT to one place within the EMUCS program, rather than duplicating the VT search code in the later cost calculation and binding phases.

Finally, initial hardware is created. The modules that are created in the initialization have no operators or regvals bound to them. Until such time as they are bound in the data path, these modules are considered to be empty, and are essentially unallocated because they have not been made part of the data path. Rather, these modules are used as seeds for the design. Initial binding costs of operators and regvals are calculated relative to these modules. Thus there must be at least one module of each type available. Since EMUCS does not allow functional units or registers to share operators or regvals of different bitwidths, an empty functional unit and register of each size needed for the design is required at the time of initialization.

6.4. PREBINDING AND MANUAL BINDING

EMUCS uses a prebinding phase to bind special operators. Three types of special operators are recognized by EMUCS: control, memory, and message operators. In addition, the user may hand choose any other operators or regvals that he wants bound to particular modules. This allows the user to partially (or fully) specify the data path if he so chooses.

6.4.1. Controller

EMUCS assumes a single controller, or sequencer, for each process for which it designs a data path. The sequencer module is created, and all control operators are bound to that sequencer module when the prebind-control-operators command is executed. The sequencer is created as a module of type "controller," with an undefined size, and no ports. The list of operators is searched, and all control operators are bound to the sequencer.

6.4.2. Memory

All array memory operators (AREAD and AWRITE) are bound to memory modules. A memory module of the appropriate bit and word size is created for each array. A memory module has an output port and a maximum of 3 input ports, numbered 2, 3, and 4, to correspond to the address, data, and bit offset inputs to the array operator, respectively. Since an AREAD operator does not have a data input, a read-only memory does not need input port 3.

6.4.3. Message Modules

The SEND and RECV message-passing operators that are used in multi-process designs as described in Chapter 3 are bound to special message modules. SEND modules are created with input ports to correspond to the operators' inputs, and output ports to match the input ports. Conversely, RECV modules are created with output ports to correspond to the operators outputs, and input ports to match the output ports. Pairs of SEND and RECV modules are connected by connecting the SEND module's outputs with the RECV module's inputs. No assumptions have been made about the implementation of the message modules. Their implementation is dependent on the target technology.

6.4.4. Manual Binding

The user may choose to both create new hardware and bind data flow items to that hardware by hand. This allows the user to superimpose his own ideas about the structure of the data path on the design created by EMUCS. When creating hardware, the user chooses the type of module he wants created (register, functional unit, multiplexor, or memory) and a size (bit width) for that module. To bind to a previously created module, the user specifies the module and the data flow operator or regval to be bound to it. Currently, regvals may be bound to registers, memory operators may be bound to memory modules, and other operators may be bound to functional units, but no other bindings are allowed. With manual binding, the user may specify all or part of the data path bindings, leaving EMUCS to fill in the appropriate interconnect between modules.

6.5. AUTOMATIC BINDING

Automatic binding is used to bind any operators and regvals that remain unbound after prebinding. In this phase, hardware is allocated and bindings are chosen automatically according to design costs.

6.5.1. Allocating New Hardware

Included in the automatic binding algorithm is a method for allocating new hardware according to design costs. An empty module of each type is maintained at all times. When costs dictate that an operator or regval be bound to an empty or unallocated functional unit or register, the functional unit or register is considered to be allocated by virtue of the binding that has made the module part of the data path. When a module is allocated, a new empty module of the same type is created so that one future binding option for any operator or register is that new hardware be allocated for it. The empty modules that are left after binding is complete are deleted in the finish-up phase.

6.5.2. The General Binding Algorithm

The calculated costs for binding operators and regvals to hardware modules are stored in two internal cost tables. One of the cost tables is dedicated to operators, the other is dedicated to regvals. The operator table has a row for each operator that needs to be bound, and a column for each functional unit available for binding. Likewise, the register table has rows for the regvals and columns for the registers. Operators cannot be bound to registers automatically. While this may be desirable for operations such as INCREMENT and SHIFT which may be implemented in registers, this option is reserved for post-processing. The general algorithm used to bind the data flow items automatically is given in Figure 6-4.

```
1.   Initialize internal cost tables
2.   While operators or regvals to bind {
3.       Find the best binding
4.       Bind the best binding
5.       Update internal cost tables
     }
```

Figure 6-4. Binding Algorithm

EMUCS uses two variations on the general automatic binding algorithm. The first variation allows the user to single-step through the binding process, making only one binding each time the command is executed. The internal cost tables are updated both before and after each binding to assure that they are correct both before the binding is made and after the binding is made. This allows the user to examine updated tables to evaluate the binding process. The second variation in automatic binding uses partial updating of the internal cost tables. When the partial update option is used, the internal tables are fully updated after each binding for the first 10% of the automatic binding process. After 10% of the operators and regvals have been bound, full updates are done less often, only after another 1% of the design has been bound. Partial updating is used after any binding that is not fully updated.

6.5.3. Updating Internal Cost Tables

Much of the CPU time spent doing automatic binding is actually spent updating the internal cost tables. As described above, two updating options are provided. The first is a full update which is precise, but slow. The second method, a partial update, may be less precise, but is faster.

Full Update. There are two approaches to fully updating the internal cost tables. Both can be quite slow. The most obvious approach uses brute force, simply updating each cost in the table. A more elegant approach searches for only those costs that need to be updated. But once a good portion of the data path has been filled in, this search may take longer than the brute force method. Currently only the brute force method is used by EMUCS.

Partial Update. If a data flow item in row *i* of an internal cost table is bound to the module in column *j*, partial updating is accomplished as given in Figure 6-5.

1. Recalculate costs of binding all unbound row operators or regvals to module *j*
2. **If** module *j* was newly allocated, recalculate costs of binding all unbound row items to new empty module
3. Update every operator and regval directly connected to item *i* in the data flow to reflect new connection costs due to the binding of item *i*

Figure 6-5. Partial Updating Algorithm

This approach updates the most obvious costs that are affected by the latest binding, but does not necessarily update all of the affected costs. therefore it is less precise than the full update, but it is much faster and still produces reasonable results.

6.5.4. Calculating Costs

Binding-costs reflect the relative amount of hardware that is needed to implement a particular binding, including any new hardware module's interconnect that must be added to the data path. The cost of binding an operator to a functional unit may include either the cost to allocate a new functional unit with the necessary functionality, or the cost to implement the necessary functionality in an existing functional unit. The cost of binding a regval to a previously unallocated register includes the cost of allocating a new register. Connection costs are considered for each binding although EMUCS does not attempt to assess connection costs for objects that have not been bound yet. Therefore, when binding an operator A + B, if neither A nor B have been bound, there are no connection costs associated with the PLUS operator. An operator or regval cannot be bound to a module that is in use at the same time that the operator or regval is in use, nor can an operator or regval be bound to a module that is part of a different partition. A module is assigned to the partition of the first object that is bound to it.

The input technology costs are described in Section 6.2.1. Figure 6-6 shows how these input costs are used to calculate binding costs for an operator. For this example, assume the following costs. Note that the first 3 costs are from the Functional Cost table while the remaining costs are from the Interconnect/Register Cost table.

- add-MINUS-function-to-empty-functional-unit (allocate a new MINUS functional unit) 200
- add-MINUS-function-to-MULT-functional-unit 1000
- add-MINUS-function-to-PLUS-functional-unit 20
- direct-connection 15
- change-input-connection-to-multiplexor 30
- add-input-to-multiplexor 25
- add-output-connection 10

Figure 6-6 shows a MINUS operator for which binding costs will be calculated, and a partial data path to which it will eventually be bound. Four registers are present in the data path, and the regvals A, B, C, D, and E have all been bound to those registers as shown. Thus all the regvals associated with the MINUS operator are already bound and these bindings will affect the cost calculation. There are 2 functional units in the partial data path, one implementing a MULT operation, the other a PLUS. There are also an empty functional unit and an empty register waiting to be allocated in the data path. Assuming no timing conflicts, the MINUS operator might be bound to any one of the 3 functional units, with varying costs to the design.

If the MINUS operator is to be bound to the MULT functional unit, several costs come into play. The first cost to consider is the cost of adding the MINUS functionality to the MULT functional unit. The high cost of adding a MINUS function to a MULT, (1000 as specified in the costs for this example) implies that the MINUS and MULT functions are not similar in the target technology. Since there are no registers connected to the MULT functional unit, connection costs must be calculated for both the inputs and the output. A direct connection can be made between register A and the multiplier's left input. The cost for this is 15. Register B can be connected to the right input by adding an output connection to register B at a cost of 10. The output of the functional unit can only be connected to register C through a multiplexor, because register C already has another functional unit connected to it. This means that the current input to register C would have to be changed to a connection to a multiplexor at a cost of 30. (Note that the change-input-to-multiplexor cost may be interpreted as the cost of allocating a new multiplexor module.) In addition to allocating a multiplexor, a second input would have to be added to the multiplexor to accommodate the output from the MULT functional unit, at a cost of 25. Therefore the total cost for binding the MINUS to the MULT functional unit is:

add-MINUS-to-MULT-functional-unit:	1000
direct-connection:	15
change-input-to-multiplexor:	30
add-input-to-multiplexor:	25
add-output-connection	10
total-binding-cost:	1080

The MINUS operator might also be bound to the PLUS functional unit. Since the PLUS and MINUS operations are similar in the target technology, it is relatively inexpensive to add the MINUS functionality to the adder. The cost of doing this is 20. While the output and right input for this functional unit are already connected to the necessary registers, the left input needs to be connected appropriately. The left input will require a multiplexor, and as described above, the costs for this are 30 and 25. Thus the cost for binding the MINUS operator to the adder is:

add-MINUS-to-PLUS-functional-unit:	20
change-input-to-multiplexor:	30
add-input-to-multiplexor:	25
---	
total-binding-cost:	75

Allocating new hardware to accommodate a binding is always an option. In the figure, the shaded functional unit represents an empty functional unit. Binding the MINUS operator to this functional unit is the equivalent of allocating a new MINUS functional unit to the design. The cost for doing this is 200. In addition to this, connection costs must be assessed. A direct connection must be made to register A, an output connection must be added to register B, and a multiplexor must be added at the input to register C. The cost for allocating a new MINUS functional unit to implement this operator is:

allocate-new-MINUS-functional-unit:	200
direct-connection:	15
add-output-connection:	10
change-input-to-multiplexor:	30
add-input-to-multiplexor:	25
---	
total-binding-cost:	280

Clearly, given the above target technology costs, the best binding for the MINUS operator is the PLUS functional unit. Changing the target technology costs, however, might lead to a different lowest cost binding. For example, if the add-MINUS-function-to-PLUS-functional-unit cost were 100, and the allocate-new-MINUS-functional unit cost were 50, it would be cheaper to allocate a new MINUS functional unit to implement the MINUS operator.

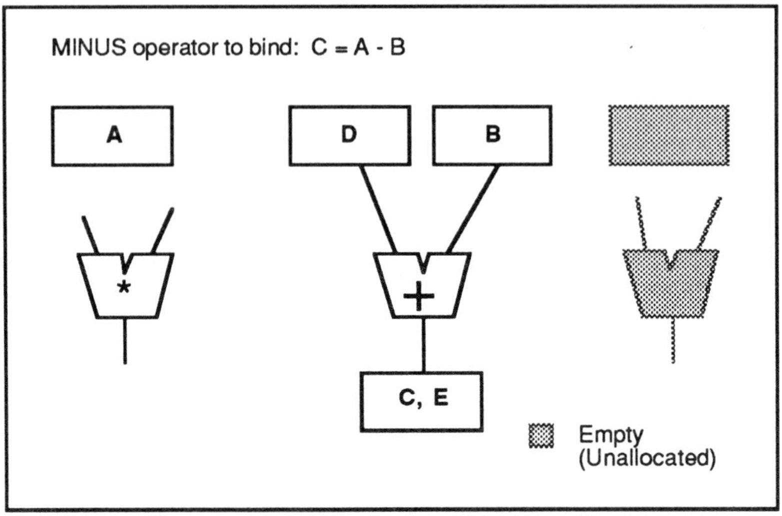

Figure 6-6. Calculating Binding Costs

6.5.5. Choosing the Next Binding

The algorithm used to choose the next binding is a simple one-step-lookahead estimation. Every possible binding is considered before a choice is made. The algorithm attempts to avoid moves that may block other good moves by considering not just the single best binding for each operator and regval, but also the second best binding because it gives a measure of the number of good binding options available for the operator or regval. This approach emphasizes the importance of binding an operator or regval that is constrained to a single good option early in the binding process. The algorithm is implemented as given in Figure 6-7.

```
Maxdiff = -1
For each operator and regval {
    low1 = lowest binding cost
    low2 = 2nd lowest binding cost
    diff = low2 - low1
    If diff is greater than Maxdiff {
        Maxdiff = diff
        best-binding = binding that results in low1
    }
}
```

Figure 6-7. Algorithm for Choosing the Best Next Binding

In this algorithm, the lowest binding costs are found by considering every possible binding for each operator and regval. The second lowest binding cost is used as an estimate of a one-step-lookahead. The difference between the two lowest costs represents the degradation in the design that may occur if that binding is not made immediately, because a future decision may block that binding. Thus, if operator A has two equally good options while operator B has only one good option, operator B will be bound before operator A because it has less flexibility.

Figure 6-8 demonstrates how a next binding is chosen. The figure shows a control step schedule and a graphical depiction of the scheduled VT. Assume the partial data path of Figure 6-6, and the costs defined previously, as well as the additional costs listed below.

- add-new-AND-functional unit 110
- add-and-to-MULT 1500
- add-AND-to-PLUS 90
- new-register 95

Both the operator and regval binding cost tables are shown in Figure 6-8. An X in the table indicates a non-option. All objects that have been bound already have X's across their rows. An X is also used when a timing conflict eliminates a binding as a possible option. The last column in each table lists the difference between the two lowest binding costs for the operator or regval. This is used to determine which operator or regval should be bound next. Since the MINUS operator has the greatest difference between its two lowest costs (*Maxdiff* = 215), it will be chosen to be bound next. For the MINUS operator, low1 = 75, so best_binding is F2, and the MINUS operator will be bound to F2, its lowest cost binding.

The choosing and binding process is repeated until all operators and regvals have been bound to data path modules.

cstep #	operations
cstep 1	E = D + B; A = E^ F * G;
cstep 2	C = A - B;

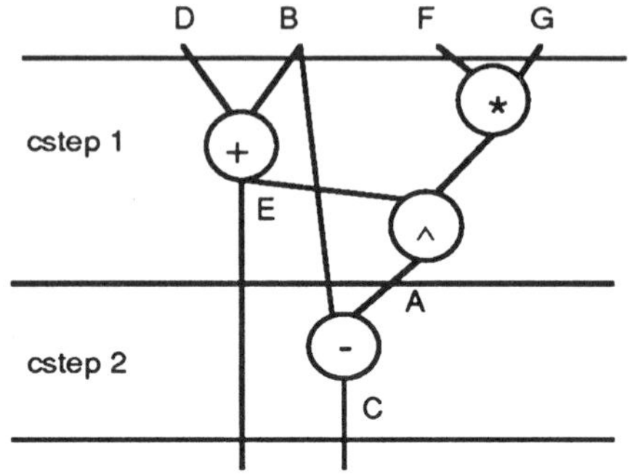

Operator Binding Cost Table

OPERATOR	F1 (*)	F2 (+)	F3 (empty)	DIFFERENCE
+	X	X	X	-
*	X	X	X	-
-	1080	75	280	215
	1530	260	240	20

Value Binding Cost Table

VALUES	R1 (A)	R2 (B)	R3 (C)	R4 (D)	R4 (EMPTY)	DIFFERENCE
A	X	X	X	X	X	-
B	X	X	X	X	X	-
C	X	X	X	X	X	-
D	X	X	X	X	X	-
E	X	X	X	X	X	-
F	15	X	100	X	110	85
G	15	X	100	X	110	85

Figure 6-8. Choosing the Next Binding

6.6. POST-PROCESSING

Several post-processing optimizations may be run to improve the final data path. Busser is a program used to choose busses and add them to the data path. Other optimizations create special purpose hardware such as increment/decrement registers and constant ROMs.

6.6.1. Busses

A post-processing program called Busser is used to choose busses for the data paths chosen by EMUCS. Busses improve the design by making them easier to lay out. A good layout is made easier by a simple interconnection scheme. Large multiplexors are difficult to lay out because they require wires from many sources to be routed through a single point. On the other hand, many modules can be connected to a single bus in a straightforward manner. Simple two-input multiplexors can be used to multiplex input from two different busses into a single destination in a simple manner. According to the partitioning guidelines described in Chapter 4, only those signals that must be passed among partitions are candidates for bussing in the Workbench.

The Busser program uses a clique partitioning algorithm to choose busses for a data path that is interconnected with multiplexors. The clique partitioning algorithm used in Busser is due to Tseng [Tseng84] and is described below. Clique partitioning is the process of dividing the nodes of a graph into a number of disjoint clusters such that each node appears in one and only one cluster, and each cluster is a complete graph. It attempts to minimize the number of busses as well as the total number of drivers and receivers.

Clique Partitioning. The clique partitioning algorithm suggested by Tseng for choosing busses considers two graphs. The first graph, G, has as its nodes all of the input ports to each module in the data path. For the data paths produced by EMUCS, any input port that is not connected to a multiplexor becomes a node (i.e. each multiplexor input port is a node, but the port that the multiplexor feeds is not a node). Edges extend between every two nodes that can share a bus (i.e. ports that are not active simultaneously unless they come from the same source). The second graph, G1, is a subgraph of G. It contains only the edges between ports that share either the same source or the same destination (i.e. they are connected through a multiplexor in the EMUCS-produced data path). This information is important because there is no apparent advantage to sharing a bus between signals that have no common end points.

Clique partitioning begins by combining those nodes from G1 that have the maximum number of common neighbors, where a node's neighbor is any other node that is connected to that node by an edge. When two nodes are combined, the graphs are reduced: in G, if a node is not connected to another node, that means that the two nodes cannot be combined because there is a timing conflict. Therefore, if a node is connected to exactly one of the two combined nodes, that node is not compatible with the combined node, so it must be deleted from both G and G1. If a node is connected to both of the combined nodes in G, it is compatible with the combined node and a single edge is maintained while the redundant edge can be deleted from both G and G1. This process is repeated until there are no edges remaining in G1.

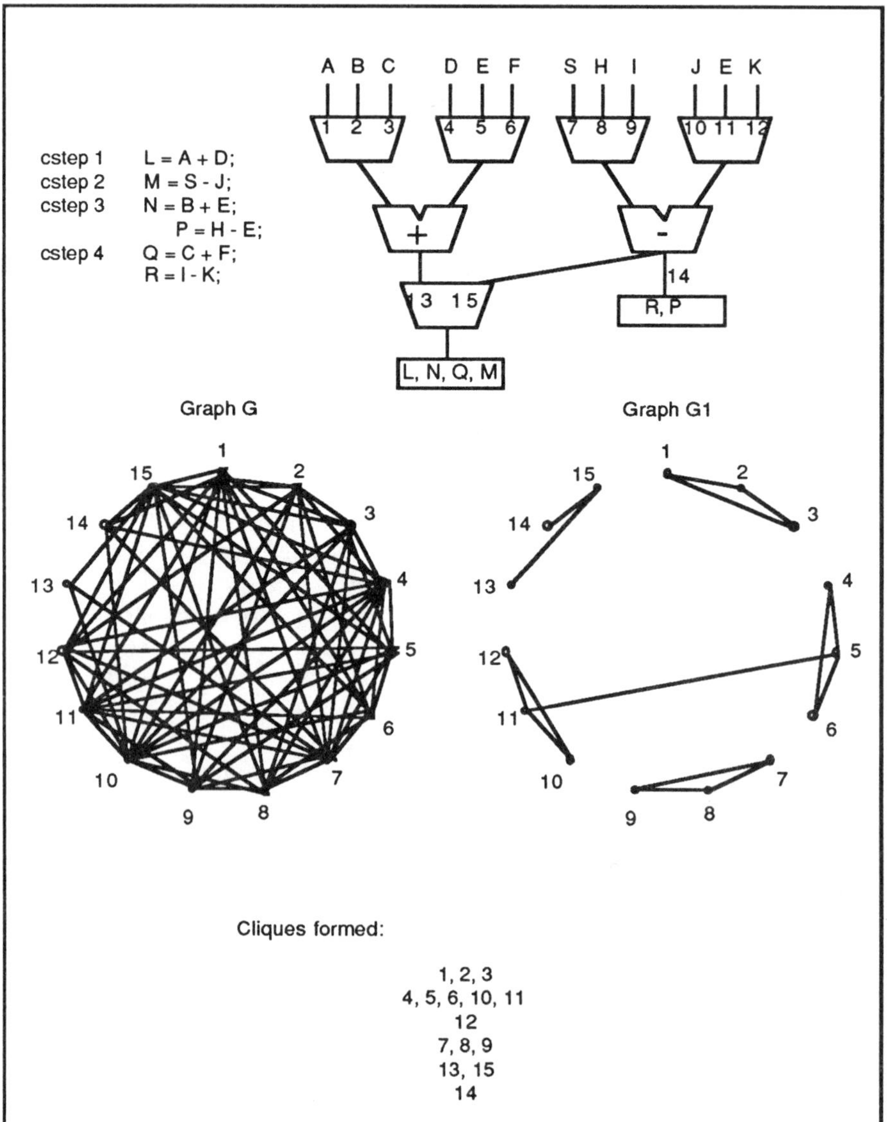

Figure 6-9. Clique Partitioning for a Simple Data Path

Figure 6-9 shows a partial data path with 15 input ports to be considered for bussing (each adder has 6 inputs connected to it by multiplexors). All wires are assumed to be candidates for bussing in this example (rather than only inter-partition connections). Node numbers are specified for each input in the data path diagram. Below the data path diagram are the graphs G and G1. Note that there are many possible bus configurations, as indicated by the high degree of connectivity in G. However, G1, the graph that indicates advantageous node groupings is quite sparse. The cliques chosen for this partial data path are listed at the bottom of Figure 6-9.

Partitioning information is further used to reduce the number of busses. By considering the partitions as high level nodes, the advantage of combining busses that service common partitions can be seen. Using this consideration, busses are combined to form "super-busses" in a second global pass of clique partitioning.

Adding Busses to the Data Path. After choosing cliques for a data path, the combined nodes are used to define busses for the design. All ports in a combined node are connected to a single bus. The algorithm is described in Figure 6-10.

```
1.  For each clique {
2.      If there are at least 2 nodes in the clique {
3.          Make a bus module
4.          For each node {
5.              Get the port associated with the node
6.              Reconnect the port through the bus
            }
        }
    }
7.  Remove any multiplexors with less than 2 inputs
```

Figure 6-10. Clique Partitioning Algorithm

The Busser program has the effect of removing from the design multiplexors with many inputs that are therefore difficult to route. Small multiplexors which for the most part get their inputs from busses and are therefore easy to route remain in the design. Figure 6-11 shows the busses for the example given in Figure 6-9.

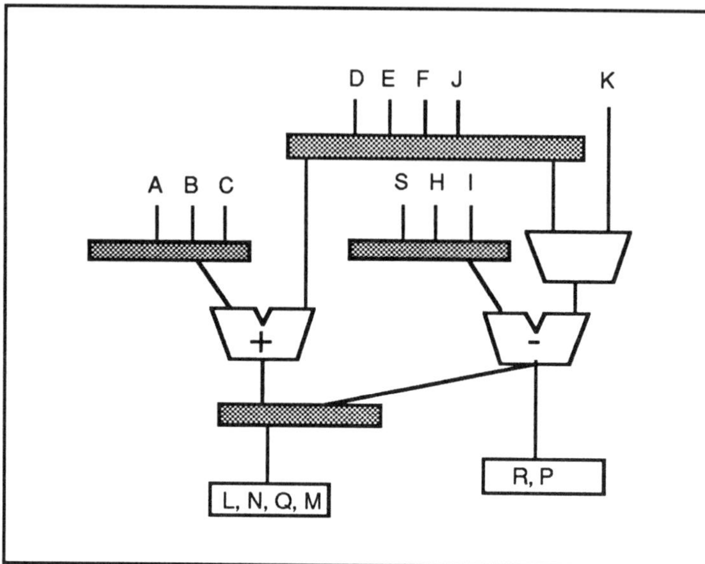

Figure 6-11. Busses Chosen by Clique Partitioning

6.6.2. Special Purpose Registers

Several operations can be implemented by registers. Those special operations that are recognized by EMUCS post-processing are: CLEAR, INC, DEC, and 1 bit shift operations. It is generally considered advantageous to implement these operations in registers when possible. Therefore, after a data path has been designed, it can be scanned for modules that can be optimized in this manner. If the user chooses this option, the appropriate registers are redefined to have the additional functionality, the functional units that originally implemented the functions are eliminated, and the appropriate connections are made.

6.6.3. Constant ROMs

Any multiplexor or bus whose inputs are exclusively constants may be transformed into a constant ROM. The transformation is a simple one. The designated multiplexor or bus and its input constants are removed from the design, and a ROM of the appropriate size is added in their place. The address port of the ROM is connected to the controller, and this port is activated whenever one of the constant values in the ROM is needed.

6.7. FINISH UP

The finish up phase is used to clean up the design and write out the data path and control schedule files. Clean up is a simple process of removing any unused ports and modules that were created during the allocation process. After the clean up has been accomplished, the data path and control schedule may be written out.

6.9. SUMMARY

EMUCS is a general purpose data path allocator that can address a broad spectrum of designs. It makes use of technology dependent input files to create a design that matches a target technology. It can make use of high level structural information provided by partitioning in creating its designs, and has post-processing steps that can improve the design. Results of using EMUCS and the other synthesis tools are given in Chapter 8.

7. Microprocessor Synthesis (SUGAR)

The four previous chapters describe one approach to Algorithmic and Register-Transfer level synthesis. In contrast to these four separate design tools that work together, the SUGAR microprocessor synthesis tool combines elements of each of these tools. This chapter will highlight three issues in register-transfer synthesis by discussing how the organization of SUGAR differs from the previous approach. The issues are: how to handle interactions between the design steps and substeps in algorithmic and register-transfer level synthesis, the differences between style-specific and general-purpose synthesis methods, and the role of knowledge representation methods.

SUGAR uses a style-specific synthesis method for the microprocessor design style. It rapidly synthesizes microprocessors by identifying features specific to them. Experimental studies show that the style-specific knowledge-rich approach can produce good designs comparable in quality to commercial designs and better than the designs generated by some general-purpose synthesis programs. Although definite conclusions cannot be drawn in favor of either the style-specific or the general-purpose methods, the comparisons are nevertheless interesting.

Several experiments were done to evaluate SUGAR. Designs for the MCS6502 and MC68000 microprocessors generated using different constraints are presented and compared with commercial designs and, where data exists, with experimental results of other synthesis programs. These results are given in Chapter 8.

In this chapter, Section 7.1 presents an overview of SUGAR. Section 7.2 describes behavioral representations and transformations to restructure behavior. Section 7.3 describes execution unit organization analysis. Code generation is described in Section 7.4 and code selection is described in Section 7.5. Section 7.6 explains how symbolic registers are assigned to structural registers and how symbolic busses are assigned to structural busses. Section 7.7 draws on the descriptions of the phases presented in earlier sections to show how the phases are organized and how they communicate with each other. Finally, Section 7.8 summarizes the chapter.

7.1. ORGANIZATION OF SUGAR

While the general synthesis path involves several tools, SUGAR is a single tool that encompasses a complete approach in several phases, which correspond roughly to the general synthesis tools. Figure 7-1 shows that the phases in SUGAR can be grouped into three major tasks, namely behavioral transformations, preliminary synthesis and synthesis. Sections 7.1.1, 7.1.2 and 7.1.3 respectively describe behavioral transformations, preliminary synthesis and synthesis. Section 7.1.4 describes the control structure of SUGAR.

7.1.1. Behavioral Transformations

SUGAR's use of behavioral transformations is very similar to their use in the general synthesis path. The purpose of behavioral transformations was described in Chapter 3. SUGAR recognizes two type of behavioral transformations: data flow and control flow. Data flow transformations alter data flow while preserving control flow and control flow transformations alter control flow while preserving data flow.

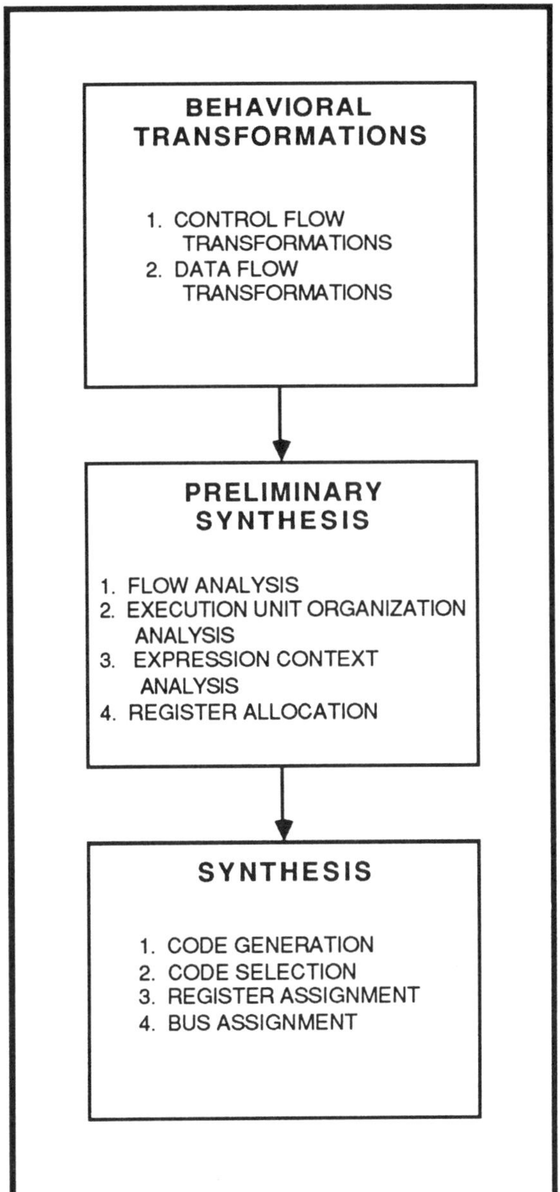

Figure 7-1. Organization of SUGAR

7.1.2. Preliminary Synthesis

The preliminary synthesis phases primarily analyze the behavior and record the results of the analysis in the behavior by annotating the internal behavioral representation. This roughly corresponds to EMUCS' initialization phase, although SUGAR is able to do a more thorough preliminary analysis by making use of microprocessor-specific information. As in the general path, the recorded information is used during synthesis to make design decisions. A few phases create data structures that will be used later by phases in the synthesis task to allocate hardware. The preliminary synthesis task consists of the following four major phases:

1. *Flow Analysis:* This phase uses classical flow analysis algorithms [Aho86] to analyze the lifetimes of variables, to determine live and dead variables, and to eliminate dead code. In the general synthesis path, dead code elimination is performed by the transformations, and lifetime analysis is performed by EMUCS.

2. *Execution Unit Organization Analysis:* This phase analyzes the behavior and calculates various parameters such as data width, address width, external data bus width and external address bus width. The relationship between these parameters is used to determine how the execution unit should be organized. Execution unit organization analysis is described in Section 7.3. In contrast, EMUCS does not perform a similar analysis. This task is necessarily specific to the microprocessor because in the general case, separate address and data sections cannot be assumed.

3. *Expression Context Analysis:* This phase analyzes each expression in the behavior and determines how it is used in relation to the expressions surrounding it. For instance, the result of an expression may be used only as a memory address. This information is recorded and used later by a synthesis phase to assign an appropriate type of register to the expression. As for Execution Unit Organization Analysis, this phase is specific to microprocessor design.

4. *Register allocation:* In the EMUCS initiation phase, the VT is searched for regvals that will have to be bound to registers.

SUGAR does a similar thing, but takes a different view of it. SUGAR's local and global register allocation phases create data structures called *symbolic registers* to record information about storage needed to hold the values of expressions. Later these symbolic registers will be merged in a step similar to EMUCS' binding phase to produce structural registers. The global register allocation phase allocates storage for variables, procedure parameters and redundant expressions while the local register allocation phase allocates storage for intermediate expressions.

7.1.3. Synthesis

The synthesis task maps the input behavioral description onto register-transfer level hardware components and control sequences. In the general synthesis path, this is done by CSTEP, EMUCS, and Busser. In SUGAR, this task consists of the following 4 major phases:

1. *Code Generation:* Code generation maps the expressions and procedures at the behavioral level to a list of uncompacted control sequences and a network of symbolic units using a rule base to select alternative ways to synthesize each behavioral expression. Symbolic busses are also generated in this phase. The rule base is selected based on the information recorded by the execution unit organization analysis phase. Code generation is described in detail in Section 7.4.

2. *Code Selection:* Code selection consists of three subtasks: scheduling, allocation and evaluation. The scheduling task packs microoperations into control states and the allocation task allocates and binds functional units. The evaluation task analyzes a design to determine its quality and uses the results of the analysis to guide design decisions. Code selection is described in detail in Section 7.5.

3. *Register Assignment:* The register assignment phase assigns symbolic registers created by the register allocation phase to structural registers. The assignment is guided by information recorded by the flow analysis phase, the expression context analysis phase and the code selection phase. Register assignment is described in detail in Section 7.6.

4. *Bus Assignment:* The bus assignment phase assigns symbolic busses created in the code generation phase structural busses. The assignment is guided by the information recorded by the execution unit organization analysis phase, the code selection phase and by the bus model chosen by the designer. Bus assignment is described in detail in Section 7.6.

7.1.4. Control Structure of SUGAR

SUGAR has a simple control structure. Each phase is identified by a name and is associated with a LISP function. The names of the phases are kept in a queue. Each phase is executed by removing its name from the queue and calling the associated LISP function. SUGAR provides support for setting break points at the beginning of phases and continuing a design from break points. The state of the design can be saved at a breakpoint with the LISP *dumplisp* function and then resumed later.

7.2. BEHAVIORAL TRANSFORMATIONS

The first phase of SUGAR translates the source description to an internal representation and applies transformations to prepare the behavior for hardware allocation. This section describes the behavioral representations and the transformations defined on them.

7.2.1. Behavioral Representation

ISPS. In ISPS, application dependent information can be attached to any part of the description using *qualifiers* — lists of arbitrary attributes, with each attribute having an optional value. Qualifiers are not interpreted by the ISPS compiler; instead, their interpretation is left to the application. For example, a qualifier, *{time: 100}*, may be attached to a loop to imply a maximum loop time of 100 time units. A specific application program may interpret this as a requirement that must be met.

Qualifiers are used to pass important information to SUGAR. The qualifiers fall into 3 categories:

- Qualifiers required for correct synthesis. Examples are architectural variable declarations. Architectural variables correspond to architectural registers. Variables which are not declared as architectural may be overlaid with other variables during register allocation or may be eliminated by the dead code eliminator.

- Qualifiers required to describe the bus controller for memory accesses. SUGAR does not synthesize the bus controller. Instead, memory accesses are described by ISPS procedures tagged with qualifiers.

- Qualifiers which suggest useful hints for good synthesis of microprocessors. Examples are identification of ISPS procedures as instructions or effective address routines.

The qualifiers used by SUGAR are all of the form:

{sugar: qualifier-name}

Figure 7-2 shows portions of a MCS6502 ISPS description augmented with qualifiers. The fetch-decode-execute loop in procedure *run* is identified by the main procedure qualifier *main*. *run* contains an instruction dependent branch *(DECODE)* which is identified by the qualifier *ibranch*. The procedure *group1* in the *DECODE* statement calls the procedure *ORA* identified as an instruction by the qualifier *instruction*. The behavioral partitioner, described in Section 7.2.2, uses the *ibranch* qualifier to determine how to build the instruction decoder to decode the ORA instruction. If the qualifier were absent, SUGAR would not be able to recognize that the branch in *run* describes instruction decoding. The transformation to in-line expand procedure calls uses the qualifier *operand* attached to the procedure *get1* to prevent the in-line expansion of the call to *get1* in the procedure *ORA*. If the qualifier were absent, all calls to *get1* would be expanded in-line. The expansion would increase the size of the control unit since after expansion it would be impossible to share the effective addressing modes described by *get1*.

TCOL. SUGAR does not translate an ISPS description directly into an internal representation. Instead, the **mill** program translates the description to a VT and the VT is then translated by a simple C program into the internal tree representation language, *TCOL* [Leverett80]. A tree representation was chosen instead of a linear representation such as the VT, since it is easier for a human to visualize a program structure by looking at a tree. This feature was useful in the development of SUGAR.

```
run {main; sugar: main} :=
BEGIN
    R = Fetch() NEXT        ! Fetch instruction
    DECODE {sugar: ibranch} IR<1:0> =>
    BEGIN
        '001 := group1(),
        '010 := group2(),
        '000 := group3(),
        '011 := UNPREDICTABLE()
    END NEXT
    RESTART RUN
    END,

group1 :=                   ! Group 1 instruction decode
BEGIN
    DECODE {sugar: ibranch} IR<7:5> =>
    BEGIN
        #0 := ora(),        ! Or Acc
        #1 := and.(),       ! And Acc
                .
                .
                .
     END
END,

ora {sugar: instruction} :=
BEGIN
    TEMP.S1 = A NEXT
    A = A OR get1() NEXT
    setcc(OR.OP, TEMP.S1, get1, A)
END,

get1() <7:0> {sugar: operand} :=
BEGIN                       ! Operand fetch for Group1
    DECODE {sugar: ibranch} IR<4:2> =>
    BEGIN
        2 := (get1 = immed()),
OTHERWISE := (get1 = BusRead(eadd()))
    END
END,
```

Figure 7-2
Portions of a MCS6502 ISPS Description Showing Qualifiers

A TCOL tree is an abstract syntax tree that resembles a parse tree with the difference that punctuation terminals in the grammar such as parentheses do not appear in the tree. Control constructs are not broken down into tests and jumps but are retained as trees. Figure 7-3 shows examples of TCOL trees.

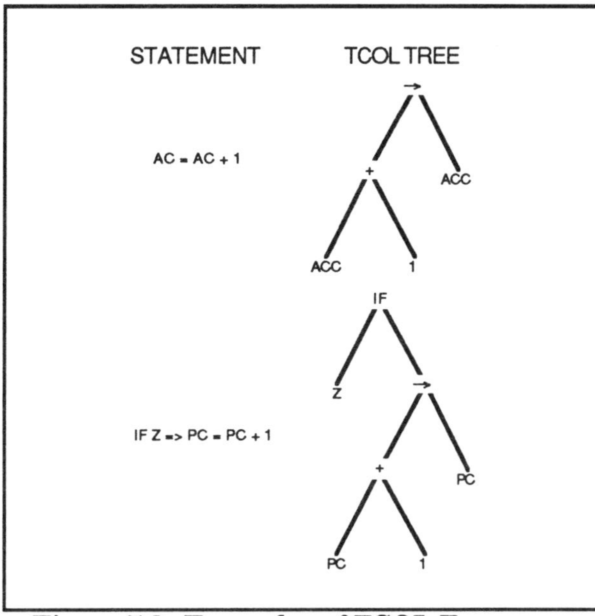

Figure 7-3. Examples of TCOL Trees

Each TCOL node consists of four slots: node type, node width, node sons and a node property list. The property list is a list of arbitrary attribute/value pairs. The property list is the primary mechanism SUGAR uses to pass information between phases.

Flow Graph. Certain phases of SUGAR, such as common subexpression analysis, context analysis and condition code analysis operate on a data flow representation. The data flow representation is a control flow graph of basic blocks like those used in many compilers. Each basic block is a simple list of tuples possibly ending in a tuple that describes a branch to another basic block. SUGAR contains facilities for translating back and forth between the TCOL and the flow graph representations.

7.2.2. Transformations

Snow [Snow78] defined several behavioral transformations to reduce the biases inherent in algorithmic languages and showed that the transformations can have a significant effect on the quality of designs. Many of those transformations, as well as others, were defined in Chapter 3 of this book, or in [Walker83].

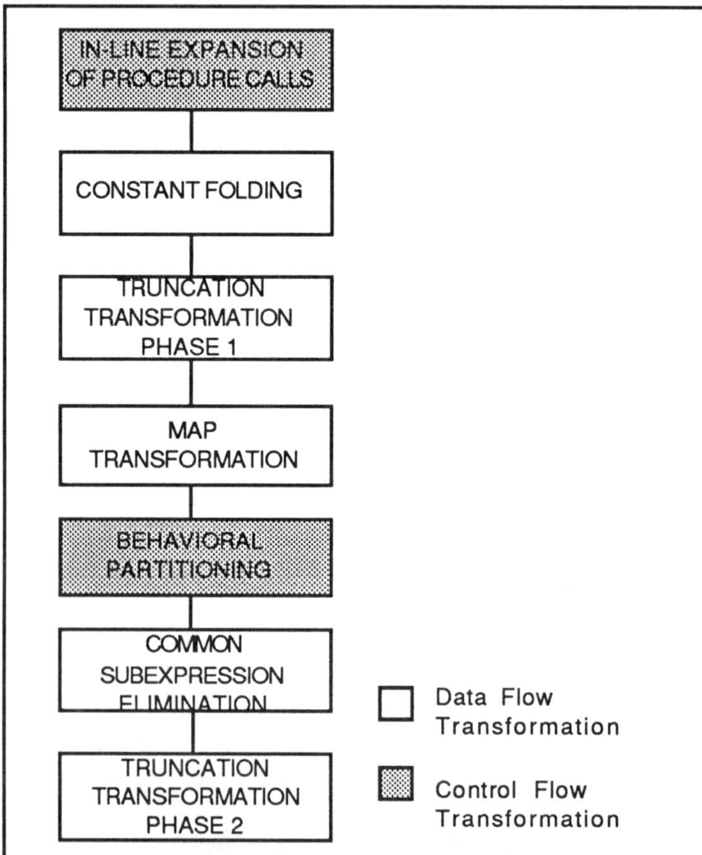

Figure 7-4. Behavioral Transformations in SUGAR

The transformations used by SUGAR fall into two main categories: control flow transformations and data flow transformations. Control flow transformations (e.g. in-line expansion of procedure calls) alter the control flow of the behavior while data flow transformations (e.g. constant folding) preserve the control flow but alter the data flow. The transformations in the Workbench's general synthesis path operate on a VT, while the transformations in SUGAR operate on TCOL trees. A small subset of the transformations defined by Snow have been implemented in SUGAR. In addition to Snow's transformations, some new transformations have been developed.

Figure 7-4 shows a flow chart of the transformations. Only the behavioral partitioning transformation is described in the remainder of this section. In-line expansion of procedure calls, constant folding and common subexpression are described in Chapter 3 or in [Walker83]. Map transformation and truncation transformation were defined by Oakley and are described in [Oakley79].

Behavioral Partitioning. SUGAR's behavioral partitioning can be compared with APARTY's architectural partitioning. Both begin with a behavioral description and find high level partitions. However, while APARTY may discover an address/data partitioning scheme, or may separate instructions through control clustering, SUGAR is able to be more precise because it was developed specifically for the microprocessor style of design. It partitions a microprocessor's behavior into separate instruction and addressing mode routines to allow parallel decoding of instructions and addressing modes.

Usually, instruction decoding is described by subdividing instructions into groups. Using nested conditionals, each group may be further subdivided into finer subgroups. Similarly, addressing mode decoding may be described using nested conditionals. The nested conditionals result in sequential decoding if the behavioral control structure is mapped directly to hardware. Normally, microprocessors use an instruction decoder to do this decoding in parallel and generate a different control sequence address for each instruction and each addressing mode. SUGAR designs such a control structure by partitioning the behavior into separate routines for each instruction and addressing mode and determining the conditions under which each routine can be executed.

To do this, SUGAR differentiates between *instruction dependent conditionals* which determine unique instructions and addressing modes and *data dependent conditionals* such as branches on processor condition codes by inspecting the qualifiers attached to the conditionals by the designer. SUGAR finds all possible execution paths that pass through instruction dependent conditionals. Each path is used to build a routine consisting of the conditions under which the path can be executed and the statements that lie on the path. When the routines are mapped to hardware, the conditions are extracted from the routines and used to design the instruction decoder so that it may generate the control sequence addresses for the routines.

An intriguing possibility is that APARTY could be used to identify the instructions for SUGAR and eliminate the need for the designer to use the *ibranch* qualifier to describe instruction dependent branches in the behavioral description.

Figure 7-5 shows part of the instruction decoding tree for the MCS6502 microprocessor and an instruction dependent path for the ORA instruction in bold. The instruction dependent path shows that the ORA instruction with INDX effective address mode is executed when IR<1:0> is 1, IR<7:5> is 0 and IR<4:2> is 0. The behavioral partitioner automatically generates all execution paths like the one shown in the example.

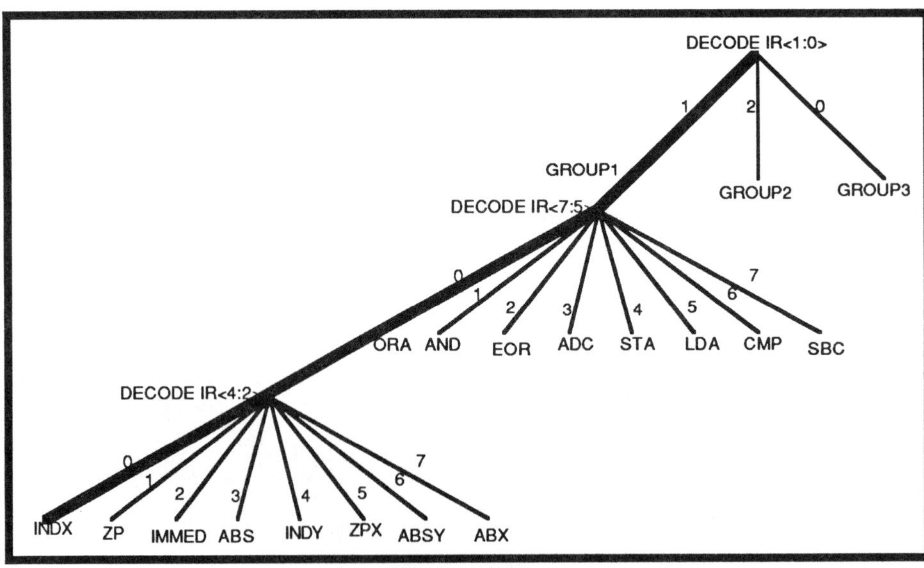

Figure 7-5
Partial Instruction Decoding Tree for MCS6502 and Path Showing Execution of ORA Instruction

Significant savings can be realized for processors with complex instruction formats since the descriptions of decoding in such processors tend to be complex. Table 7-1 shows the effect of behavioral partitioning and in-line expansion of procedure calls on the execution time of the Motorola MC68000 Add Address Register Indirect instruction. The first column shows the number of decodes in the instruction. The second column shows the number of procedure calls in the instruction. The third column shows the estimated number of data-flow cycles required to decode and execute the instruction. The first row shows that before partitioning the instruction has 8 decodes and 9 procedure calls and requires 28 cycles to execute. The second row shows that after partitioning and in-line expansion of procedure calls, the instruction has only 2 decodes and no procedure calls and the execution time drops to 3 cycles. The last row shows that the commercial processor's execution time of 2 cycles is better than SUGAR's execution time by only 1 cycle. This discrepancy is due to differences in the prefetch schemes.

Table 7-1
Effects of Control Transformations on MC68000 Add An@ Execution Times

Description	Decodes	Calls	Cycles
Before Partitioning	8	9	28
After Partitioning	2	0	3
Commercial MC68000	*	*	2

The behavioral partitioning method described above was influenced by Oakley's [Oakley79] work on the automatic generation of machine tables for a compiler-compiler from behavioral descriptions. Oakley, however, uses a complex symbolic execution method to determine the conditions for the execution of instructions and addressing modes.

7.3. EXECUTION UNIT ORGANIZATION ANALYSIS

Interconnections on a VLSI chip often occupy a large portion of the chip area. Choice of a poor interconnection structure can lead to poor use of chip area and impractical layouts. A weakness of many existing synthesis programs is that they design interconnections bottom-up. As a result, they often produce either highly distributed designs with too many multiplexors and too few busses, or highly bus oriented designs with too many busses. To prevent such a situation from occurring and to ensure good floor plans, SUGAR selects the execution unit organization early in the design.

Current technology limits the number of global busses to only two or three. Most contemporary commercial microprocessors use two busses. The choice of a two bus scheme has the advantage that dense two-output register cells can be used. This allows busses to be routed directly over functional units and registers in the execution unit because each register can be connected to both busses. The current implementation of SUGAR only understands 2-bus execution units.

Besides choosing the number busses, SUGAR also selects a suitable bus model. A bus model describes the arrangement of processing sections in the data path. Low performance processors normally have only one processing section that does both address and data computations. Complex processors such as the IBM Micro 370 have separate address and data sections so that address and data computations can be done in parallel. The 2-bus, 2-section model in part *b.)* of Figure 7-6 shows the organization of the processing sections in the IBM Micro 370 style bus model.[1] Sections may be further subdivided into subsections. For example, the Motorola MC68000 has 16-bit data paths. It has a 16-bit data processing section and a 32-bit address processing section that consists of two 16-bit subsections. The 2-bus, 3-section model in Figure 7-6 part *c.)* shows the organization of the processing sections in the Motorola MC68000.

SUGAR determines the number of processing sections in the execution unit based on the following two factors:

- The relationship between the widths of memory data bus, address variables, data variables and instructions.

- The widths of the most frequent address and data computations and the length of the most frequent instruction.

For example, the Motorola MC68000 behavior describes a 16-bit memory, 32-bit data registers and 32-bit address registers. Since the instructions are all multiples of 16-bits, the data path should be at least

---

[1]Also referred to as the IBM μ370 bus model.

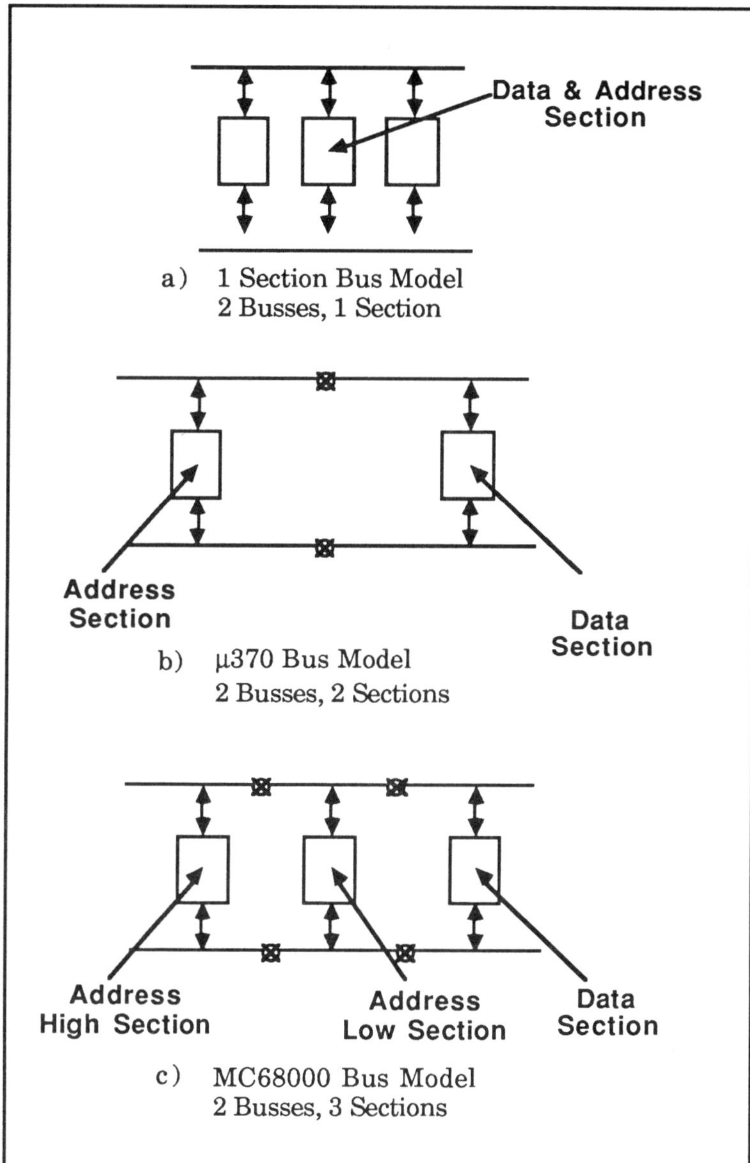

Figure 7-6. Bus Models

16 bits wide. The most frequent effective address computations are 32-bits wide, but the most frequent data computations are only 16-bits. Therefore, SUGAR selects a 16-bit data section and a 32-bit address section. To ensure a good topology, the 32-bit address section is divided into adjacent 16-bit address-low and 16-bit address-high sections. All the sections share the two busses. The busses have bidirectional bus couplers at the section boundaries. This allows the execution unit to be dynamically partitioned into three independent sections enabling up to six parallel bus transfers.

Based on the execution unit organization analysis, SUGAR determines which bus models are feasible for the given behavior. The designer is then asked to select a desirable bus model at the beginning of the code generation phase. The IBM Micro 370 and the Motorola MC68000 bus models have been implemented in SUGAR.

Later phases of SUGAR use model-specific knowledge in code generation, allocation, scheduling, register assignment and bus assignment. For instance, the code selector allocates and schedules MC68000 style models in such a way that computations are done locally in appropriate sections of the execution unit and register transfers between sections are minimized to ensure high parallelism.

7.4. CODE GENERATION

Code generation maps the expressions and procedures at the algorithmic level to a list of uncompacted control sequences and a network of symbolic units. Most synthesis programs, including the Workbench's general synthesis path, do not have a distinct code generation phase since they use a simple method to map algorithmic level operators to hardware functional units. Specifically, each algorithmic level operator is allocated a hardware unit of the same type and size as the operator on the assumption that there is only one way to synthesize the operator. In practice, this assumption is too restrictive since many operators can normally be synthesized in different ways. Each of these ways is described by a distinct control sequence. For instance, a 16-bit addition operator can be synthesized using a 16-bit adder or an 8-bit adder. In the first instance, the 16 bits are added in parallel while in the second instance, only 8 bits are added in parallel, requiring 2 passes through the adder to complete the 16 bit addition. In addition to the simple one to many mapping, it may also be possible to synthesize efficiently complex expressions containing several operators using special hardware networks.

In contrast to synthesis programs, software compilers normally have a distinct code generation phase since an architecture's rich instruction set often makes it difficult to decide what is the best way to generate machine instructions for an expression. Code generation is usually driven by a library of code selection *templates*. Each template consists of a pattern tree and a code sequence. The code sequence is emitted when a subtree in the intermediate representation matches the pattern tree.

If several templates match an expression, a method is needed to select a desirable template. One possibility is to select a template based on local features, but local code selection normally leads to poor code sequences since microprocessors have low-level parallelism, which makes it difficult to determine the quality of code for an expression without considering its relationship to the surrounding code. Another possibility is to generate all feasible code options and then select a desirable code option based on some form of global analysis. Vegdahl [Vegdahl82] uses this method to generate code for retargetable microcode compilers and this is also the method used in SUGAR. Code options are represented by *AND/OR* code trees. In an *AND/OR* tree, each node is either an *AND* node, an *OR* node or a *LEAF* node. Each *LEAF* node consists of uncompacted control sequences. For each routine, all possible code options are built by generating all possible paths through the *AND/OR* code tree for that routine.

This section describes the code generation algorithm. It begins with a description of templates in Section 7.4.1. Section 7.4.2 details the algorithm and the algorithm is illustrated with an example in Section 7.4.3.

7.4.1. Templates

The code generation algorithm generates code by using a local pattern matching technique to match a library of templates against subtrees in the behavioral description. There is one template library for each type of execution unit, where the type of the execution unit is determined by the number of processing sections in it.

Each template consists of a condition or left-hand side (*lhs*) and a result sequence or right hand side (*rhs*). The lhs is a pattern tree. Nodes of a template's pattern may specify what restrictions a matching behavioral subtree must satisfy for the template to be selected. The restrictions are described by arbitrary LISP expressions called *predicates*. The leaves of a pattern tree are variables representing the pattern's operands. The leaves may be constants or partially specified locations. A partially specified location may be used to describe features that a matching behavioral subtree should satisfy. For instance, a partially specified location may restrict the locations of matching behavioral subtrees to certain processing sections of the micromachine. A template's rhs describes what microoperations are to be generated, their timing and the data dependencies between them. In addition to the predicates associated with the individual nodes of a pattern tree, each template can have another arbitrary LISP predicate called the *template predicate*. A template is said to match only if the lhs pattern matches and the template predicate is satisfied.

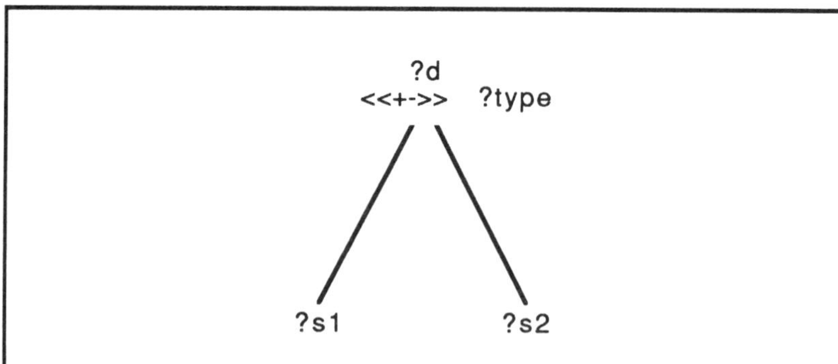


```
control state:
        microoperations:
                (-> ?s1{1} bus1{1 2 3} alutn)
                (-> ?s2{1} bus2{1 2 3} alutn)
                (function ?type alutn{2 3 4})
                (-> (alutn o){4} alutno{4})
control state:
        microoperations:
                (-> alutno{1} bus3{1 2 3} ?d{3})
```

Figure 7-7
Example of a Template — Pattern and Code Sequence

Figure 7-7 shows a template for addition and subtraction in the *a* (address) context. For the sake of clarity, details have been omitted from the figure. The actual templates are described in LISP. The root of the lhs pattern tree can be a + operator or - operator. A *?* indicates a variable to which the matched node will be bound. There are no restrictions on the locations of the operands. The rhs code sequence describes how the operands represented by the variables *?s1* and *?s2* are transferred via symbolic busses *bus1* and *bus2* to a symbolic arithmetic-logic unit alutn and then transferred from the symbolic output latch *alutno* of *alutn* to the destination *?d* via the symbolic bus *bus3*. The integers in braces are clock phases used to describe the timing of the microoperations. For instance, operand *?s1* is read during clock phase 1, data is transferred on the bus *bus1* during phases 1, 2 and 3. *alutn* computes during clock phases 2, 3 and 4. *alutno* is latched during clock phase 4 from the output port *o* of *alutn*. *alutno* is transferred during the next control state to *?d*.

7.4.2. Algorithm

The code generator uses a *Maximal Munching Algorithm* (MMM) [Cattell82] algorithm for pattern matching. MMM attempts to match the largest possible behavior subtree at each stage of the pattern matching. The lhs pattern tree of each template in the given template library is matched top-down against each behavioral tree in each behavioral routine. If the template matches the behavioral tree, the template's rhs is *fired* to generate the code sequence. If more than one template matches a behavioral tree, an *OR* node is created to describe the choice. The matched operands of a behavioral tree are matched recursively. Code generated for different parts of the behavioral tree are combined using *AND* nodes.

The code generation method described above combines the algorithmic and knowledge-based approaches. The pattern matcher is an algorithm while the templates are rules. Since the pattern matcher and the templates are independent of each other, new templates can be added incrementally without making any changes to the pattern matcher.

7.4.3. An Example

Consider a behavioral tree from the MCS6502 microprocessor behavior shown in Figure 7-8. The tree describes the concatenation (@)

of an 8-bit constant *1* with the result of an 8-bit increment of the stack pointer, *S*. The figure shows the contexts associated with each node of the behavioral tree.[2] For instance, the + node occurs in the *al* (address low) context. Sometimes, nodes of a pattern tree restrict matching behavioral nodes to certain contexts. The figure also shows the locations for the leaves of the trees. The locations describe symbolic registers allocated by the register allocation phase before code generation. For instance, the leaf operand of the + is associated with the location *(S al)*. Here, *al* specifies that the symbolic register *S* resides in the *al* section of the micromachine. The @ node is associated with the location *(tn1 a)*. Here *a* represents the address section.

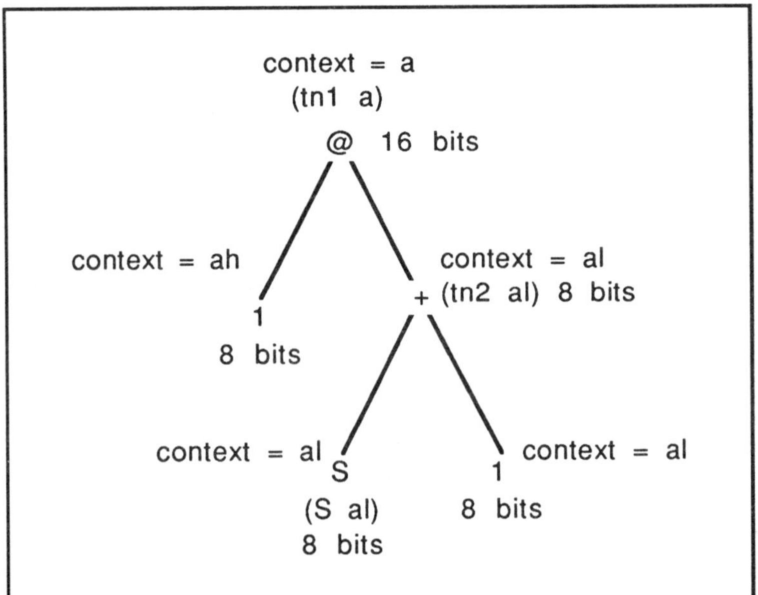

Figure 7-8. Behavior Tree for Code Generation Example

[2] The contexts were determined by the context analysis phase before code generation.

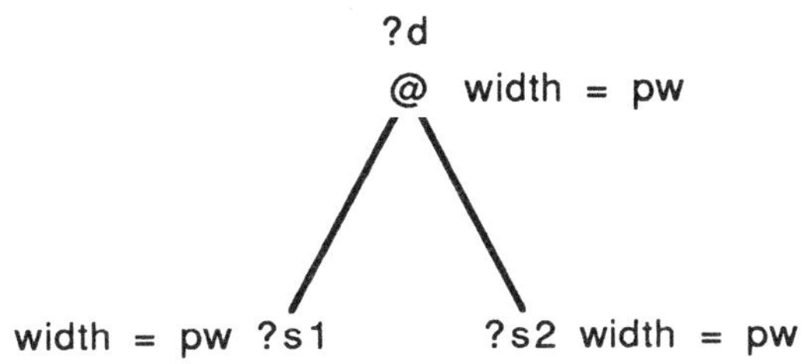


```
state = (make-state)
(emit-uOp state (-> ?s1 bus1 ?d<aw-1:pw>))
(emit-uOp state (-> ?s2 bus2 ?d<pw-1:0>))
(emit-state state)
```

Figure 7-9
Concatenation Template — Pattern and Code Sequence

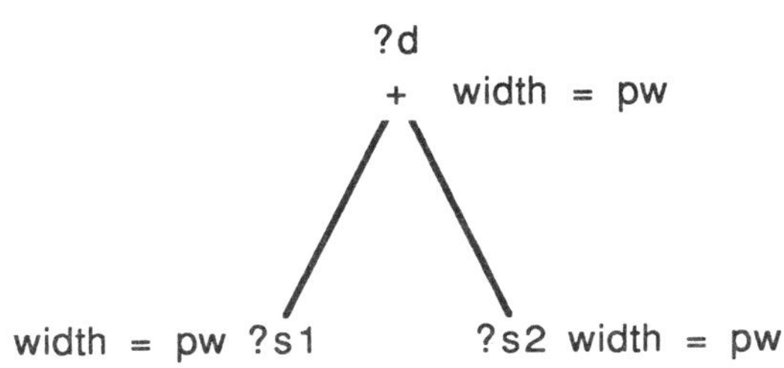


```
alutn = (make-alu pw)
alutno = (make-register pw)
state1 = (make-state)
(emit-uOp state1 (-> ?s1 (make-bus pw) (alutn ia)))
(emit-uOp state1 (-> ?s2 (make-bus pw) (alutn ib)))
(emit-uOp state1 (-> 0 (alutn ci)))
(emit-uOp state1 (function (A+B) alutn))
(emit-uOp state1 (-> (alutn o) alutno))
(emit-state state1)
state2 = (make-state)
(emit-uOp state2 (-> alutno (make-bus pw) ?d))
(emit-state state1)
```

Figure 7-10
Addition Template — Pattern and Code Sequence

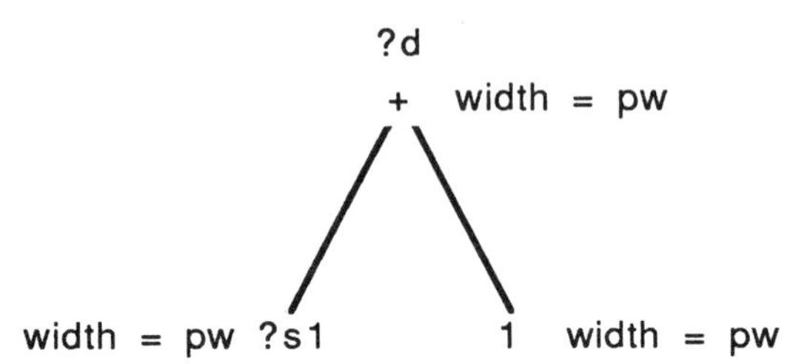


```
inctn = (make-inc pw)
inctno = (make-register pw)
state1 = (make-state)
(emit-uOp state1 (-> ?s1 (make-bus pw) (inctn i)))
(emit-uOp state1 (-> 0 (inctn ci)))
(emit-uOp state1 (function (inc) inctn))
(emit-uOp state1 (-> (inctn o) inctno))
(emit-state state1)
state2 = (make-state)
(emit-uOp state2 (-> inctno (make-bus pw) ?d))
(emit-state state2)
```

Figure 7-11
Increment Template — Pattern and Code Sequence

Assume that the template library has 3 templates as shown in the Figures 7-9, 7-10 and 7-11. The first template describes how to concatenate 2 operands, the second template describes how to add two operands using an arithmetic-logic unit and the third template describes how to increment an operand using an incrementer. In the templates, *aw* represents the address section width and *dw* represents the data path width. For the MCS6502 microprocessor, *aw* is 16 bits and *dw* is 8 bits.

For the behavior tree in Figure 7-8, the code generator first begins by generating an *OR* node *OR1* for the behavioral tree as shown in Figure 7-12. The code generator then searches for a template that matches the @ node in the behavioral tree and finds the template shown in Figure 7-9. The root of the template's pattern is pattern matched with the root of behavioral tree. The two nodes match since they are of the same type and have the same widths.

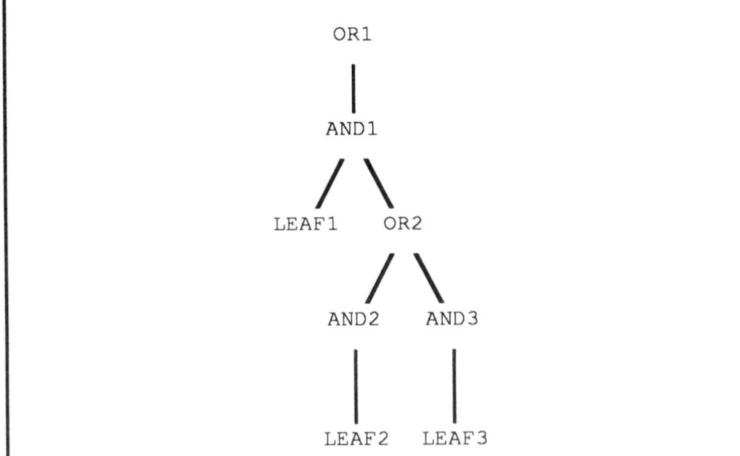


```
LEAF1
        state1:
                (-> 1 tnbus1 tn1<15:8>)
                (-> tn2 tnbus2 tn1<7:0>)
LEAF2
        state2:
                (-> S tnbus3 (tnalu1 ia))
                (-> 1 tnbus4 (tnalu1 ib))
                (-> 0 (tnalu1 ic))
                (function (A+B) tnalu1)
                (-> (tnalu1 o) tnalu1o)
        state3:
                (-> tnalu1o tnbus5 tn2)
LEAF3
        state4:
                (-> S tnbus6 (tninc1 i))
                (-> 0 (tninc2 ci))
                (function (inc) tninc2)
                (-> (tninc1 o) tninc1o)
        state5:
                (-> tninc1o tnbus7 tn2)
```

Figure 7-12. Generated Code Tree

The pattern matcher then recursively matches the pattern's operands with the behavioral node's operands. The left operand of the pattern describes an arbitrary operand; so, it matches the left operand of the @ node in the behavioral tree. The right operand of the pattern also describes an arbitrary operand; so, it matches the right operand of the @

node in the behavioral tree. Since the template matches, the template is fired and an *AND* node *AND1* is generated and a *LEAF* node *LEAF1* describing the rhs code sequence is added to the sons of the *AND1*. No other templates match the @ behavioral node.

The code generator then moves to the operands of the @ behavioral node. No code is generated at the left operand since it is a leaf. Next the code generator searches for a template for the + node in the behavioral tree. It generates an *OR* node *OR2* for the behavioral node and adds it to the sons of *AND1*.

The first template found is again the template show in Figure 7-9. The template does not match since the type of the behavioral node does not match the type of pattern node. Next, the code generator finds the template shown in Figure 7-10. This template matches the behavioral node. So, an *AND* node *AND2* is generated and the code sequence specified in the template is added as a *LEAF* node *LEAF2* to the sons of *AND2*. No code is generated at either operand of the behavioral node since they are both leaves.

The behavioral node is then matched against the next template. The template shown in Figure 7-11 matches the behavioral node. So nodes *AND3* and *LEAF3* are generated, *LEAF3* is added to the sons of *AND3* and *AND3* is added to the sons of *OR2*.

The complete code tree is shown in Figure 7-12. The leaves of the code tree are shown in detail in the figure. The locations in the microoperations shown are all symbolic registers, symbolic functional units or symbolic busses. Observe that the code generator generates code in the reverse order of execution, i.e., code is generated for the operands of an expression after code has been generated for the root of the expression.

7.4.4. Observations About Code Generation

The *AND/OR* code generation postpones the decision of how to synthesize each behavioral expression by separating the feasibility analysis from the desirability analysis. The division of labor simplifies the design of the synthesis program. The method is also appealing because the code generation and code selection phases can be designed independently with a clean interface between them.

In SUGAR, the *AND/OR* code trees constitute the interface between the two phases. The *AND/OR* code trees can be used to explore the design space extensively as explained in Section 7.5. Such an extensive exploration of the design space is not done with most synthesis programs, including the general synthesis path in the Workbench; they synthesize each behavioral expression in only one way. The *AND/OR* code generation method requires significant space for representation of the *AND/OR* code trees. A saving grace is that there are normally only a few ways to synthesize an expression. Therefore, the *AND/OR* code tree for an expression in usually small.

7.5. CODE SELECTION

Code selection consists of three tasks: scheduling, allocation and evaluation. The *scheduling task* packs microoperations into control states subject to the constraint that some criteria are satisfied. The *allocation task* allocates functional units, storage units and interconnection units, subject to the constraint that some criteria are satisfied. Allocation may be divided into a *symbolic unit creation task* and a *symbolic unit to structural unit assignment task*. The *evaluation task* analyzes a design to determine its quality.

Normally, the goal in code selection is to allocate hardware units such that the total hardware cost meets a cost constraint and the schedules meet a performance constraint. Recall from the previous section, that SUGAR generates code before selecting code. For each routine, the code generation phase passes an *AND/OR* code tree representing the different code options to the code selection phase.

Briefly, the code selection phase selects the most desirable implementation for each routine using a stepwise pruning method to discard undesirable code options from the *AND/OR* trees. The pruning is guided by evaluations of schedules generated by the code scheduler. Although allocation and code scheduling are not interleaved, the two tasks communicate by exchanging information. After a design has been synthesized, the code evaluator evaluates the design and categorizes it along several dimensions. The user can use the evaluations to refine the design and explore the design space by altering the constraints.

Code scheduling is done in three passes. In the first pass, all possible uncompacted control sequences are generated for each

AND / OR code tree. Each control sequence represents a code option. Code options which violate functional unit constraints and system cost constraints are eliminated. At the beginning of the second and third passes, a technique called *data forwarding* is used to shorten data dependency chains.

In the second pass, the code scheduler first eliminates all code options whose mean function concurrency is less than the maximum mean function concurrency measured over all code options. The *mean functional concurrency* is a measure of the average number of functional units concurrently active in a control state of a given control sequence. The selection of desirable code options is based on the heuristic that an uncompacted control sequence with a larger mean function concurrency can be expected to produced a faster compacted control sequence than another uncompacted control sequence with a smaller mean function concurrency. The code scheduler then designs the fastest schedule for each code option subject to the constraints that each schedule does not require more than a given number of functional units of each type and the schedule is feasible for the chosen bus model.

The validity of the schedules are checked during scheduling using *hardware conflict analysis* algorithms. The restrictions on how many functional units of each type are allowed is specified by the user as *constraints*. If a bus model is selected by the user before code generation, the global interconnections and execution unit topology are fixed before code generation begins. In such cases, not all schedules are feasible since some schedules may require interconnections not found in the selected bus model. If the bus model is not fixed, the interconnections are designed bottom-up after functional units and registers have been created and assigned. In such instances, all schedules are assumed to be feasible.

No hardware is allocated or assigned in the second pass. The schedules designed in the second pass are used to guide the assignment of functional units in the third pass. The code scheduler then designs a more accurate schedule based on the assignment of the functional units.

Symbolic registers are assigned to structural registers and symbolic busses are assigned to structural busses in separate phases following code selection as described in Section 7.6. Details of the code scheduling algorithm can be found in [Rajan88].

7.5.1. An Example

This section illustrates code selection with an example. Before presenting the example, the notation used in the following figures to describe symbolic hardware units is explained and illustrated below with examples.

- *inctn1* refers to an 8-bit unit. Unless a width is specified explicitly, the width should be assumed to be 8 bits.

- *(PC_r7 8)* refers to the bits *<7:0>* of the unit *PC_r7*.

- *(inctn0 i)* refers to the 8-bit port *i* of the unit *inctn0*.

Consider code selection for the Zero Page addressing mode in the MCS6502 microprocessor. Assume the design is to be synthesized using a fixed MC68000 style bus model. For the sake of clarity, symbolic busses used in register transfers and timing details have been omitted from the figures describing the code selection.

Figure 7-13 shows the user's constraints for the example. The constraints state that the design should have at most 1 8-bit arithmetic-logic unit *(alu)* in processing section *dl*, at most 1 8-bit incrementer *(inc)* in processing section *al* and at most 1 8-bit incrementer in processing section *ah*. Functional units *inc1* in processing section *al*, *inc2* in processing section *ah* and *alu1* in processing section *dl* are created before code selection. The constraints also state that the total cost of the design should not exceed 22 cost units. Assume also a maximum parallelism or mean function concurrency of 2.0.

The Zero Page addressing mode fetches a byte from the instruction stream, increments the program counter *PC_r7* to point to the next location in the instruction stream, zero extends the fetched data to 16 bits to form the effective address and fetches the operand at that address from the memory.

Figure 7-14 shows the *AND/OR* code tree for the addressing mode. Figures 7-15, 7-16 and 7-17 respectively show the initial uncompacted control sequences for the three code options *option1*, *option2* and *option3*

```
(make-constraint
    type 'alu
    width 8
    max '(dl 1)
    cost 10)

(make-constraint
    type 'inc
    width 8
    max '(al 1 ah 1)
    cost 6)

(make-constraint
    type 'system
    cost 22)
```

Figure 7-13
Constraints for Code Selection Example

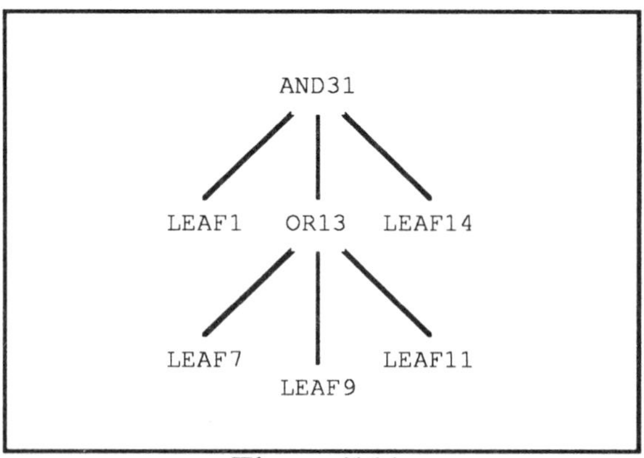

Figure 7-14
***AND/OR* Code Tree for MCS6502 Zero Page Addressing Mode**

```
LEAF1                                          ; read a byte from memory
    state0
            (memory (busreadbegin) M_r2)
            (-> (PC_r7 8) (ao 8))
            (-> (PC_r7 8 8) (ao 8 8))
            (-> (ao 8) (eab 8))
            (-> (ao 8 8) (eab 8 8))
```

Figure 7-15. Initial Uncompacted Control Sequence for Code Option *option1*

```
    state1
            (memory (busreadcontinue) M_r2)
            (-> edb di)
    state2
            (-> di tn284)
    state3
            (-> tn284 tn146)
LEAF7                                           ; increment PC
    state4
            (-> (PC_r7 8) (inctn1 i))           ; inc1 <- PC<7:0>
            (-> 0 (inctn1 ci 1))                ; carry in of inc1 <- 0
            (function (inc) inctn1)             ; inc1++
            (-> (inctn1 o) inctn1o)
            (-> (inctn1 co 1) (inctn0 ci 1)) ; carry in of inc0 <- carry
                    ; out of inc1
            (-> (PC_r7 8 8) (inctn0 i))         ; inc0 <- PC<15:8>
            (function (inc) inctn0)             ; inc0++
            (-> (inctn0 o) inctn0o)
    state5
            (-> inctn1o (tn285 8))              ; transfer incremented value
                                                ; to temp register
            (-> inctn0o (tn285 8 8))
LEAF14
    state12                          ; zero extend byte to 16
                                                ; bits
            (-> (tn285 8) (PC_r7 8))
    state13
            (-> (tn285 8 8) (PC_r7 8 8))
    state17
            (-> tn146 (tn264 8))
            (-> 0 (tn264 8 8))
    state14
            (memory (busreadbegin) M_r2)        ; fetch operand from
                                                ; effective address
            (-> (tn264 8) (ao 8))
            (-> (tn264 8 8) (ao 8 8))
            (-> (ao 8) (eab 8))
            (-> (ao 8 8) (eab 8 8))
    state15
            (memory (busreadcontinue) M_r2)
            (-> edb di)
    state16
            (-> di tn145)
    state18
            (-> tn145 tn3)
    state19
            (jmap (ipla2) ire0)
```

Figure 7-15. Initial Uncompacted Control Sequence for Code Option *option1* (continued)

```
LEAF1                                        ; read a byte from memory
          .
          .
          .
LEAF9                                                  ; increment PC
    state6
         (-> (PC_r7 8) (alutn0 ia))   ; input a of alu0 <- PC <7:0>
         (-> 1 (alutn0 ib))           ; input b of alu0 <- 1
         (-> 0 (alutn0 ci 1))         ; carry in of alu0 <- 0
         (function (A+B) (alutn0 function 1))         ; Pc + 1 + 0
         (-> (alutn0 o) alutn0o)
         (-> (alutn0 cnvz 4) alutn0cnvz)      ; save status bits
    state7
         (-> alutn0o (tn285 8))       ; transfer sum to temporary register
    state8
         (-> (PC_r7 8 8) (alutn1 ia)) ; input a of alu1 <- PC<15:0>
         (-> 0 (alutn1 ib))           ; input b of alu1 <- 0
         (-> (alutn0cnvz 3 1) (alutn1 ci 1))  ; carry in of alu1 <- carry
                                              ; out (bit 3 of status)
         (function (A+B+C) (alutn1 function 1))       ; PC + 0 + carry in
         (-> (alutn1 o) alutn1o)
         (-> (alutn1 cnvz 1 3) (alutn1cnvz 1 3))
         (-> (clogic (boole 1 (alutn0cnvz 1) (alutn1 cnvz 1)))
(alutn1cnvz 1))                       ; save status bits
    state9
         (-> alutn1o (tn285 8 8))
LEAF14
          .
          .
          .
```

Figure 7-16
Initial Uncompacted Control Sequence for Code Option *option2*

corresponding to the three leaves, *LEAF7* , *LEAF9* and *LEAF11* of the node *OR13*. The 3 control sequences differ only in how *PC_r7* is incremented. In Figure 7-15, *PC_r7* is incremented using 2 8-bit incrementers (*LEAF7*); in Figure 7-16, *PC_r7* is incremented using an 8-bit arithmetic-logic unit (*LEAF9*) and in Figure 7-17, *PC_r7* is incremented using 2 8-bit arithmetic-logic units (*LEAF11*). Note that although both LEAF9 and LEAF11 name 2 symbolic ALUs in their code, LEAF11 requires 2 actual ALUs because there is a direct transfer of data from the carry out of one to the carry in of the other, while in LEAF9 the status information is stored across state boundaries.

The first pass of code evaluation finds that *option1* requires at least 2 8-bit incrementers since *state4* uses 2 8-bit incrementers in parallel. The minimum cost of the functional units required for *option1* is 12 cost units each 8-bit incrementer costs 6 cost units. The code evaluator does not eliminate *option1* since its cost does not violate the maximum permissible system cost of 22 cost units and the constraints allow 2 incrementers to be used. The second pass of code evaluation finds that the mean function concurrency of the uncompacted control sequence for *option1* is 2.0.

```
LEAF1
            .
            .
            .
    LEAF11
    state10
        (-> (PC_r7 8) (alutn2 ia))          ; input a of alu2 <- PC<7:0>
        (-> 1 (alutn2 ib))                  ; input b of alu2 <- 1
        (-> 0 (alutn2 ci 1))                ; carry in of alu2 <- 0
        (function (A+B) (alutn2 function 1)) ; PC + 1 + 0
        (-> (alutn2 o) alutn2o)
        (-> (alutn2 cnvz 3 1) (alutn3 ci 1)) ; carry in of alu3 <-
                                             ; carry out of alu2
        (-> (PC_r7 8 8) (alutn3 ia))        ; input a of alu3 <-
PC<15:0>
        (-> 0 (alutn3 ib))                  ; input b of alu3 <- 0
        (function (A+B+C) (alutn3 function 1))      ; PC + 0 + carry in
        (-> (clogic (boole 1 (alutn3 cnvz 1) (alutn2 cnvz 1)))
        (alutn3cnvz 1))
        (-> (alutn3 o) alutn3o)
        (-> (alutn3 cnvz 1 3) (alutn3cnvz 1 3))     ; save status
    state11
        (-> alutn2o (tn285 8))
        (-> alutn3o (tn285 8 8))
LEAF14
            .
            .
            .
```

Figure 7-17. Initial Uncompacted Control Sequence for Code Option *option3*

```
state43
     (memory (busreadbegin) M_r2)
     (-> (PC_r7 8) ebus1 (ao 8))
     (-> (PC_r7 8 8) ebus1 (ao 8 8))
     (-> (ao 8) (eab 8))
     (-> (ao 8 8) (eab 8 8))
     (-> (PC_r7 8) ebus1 (inc1 i))
     (-> 0 (inc1 ci 1))
     (function (inc) inc1)
     (-> (inc1 o) inc1o)
     (-> (inc1 co 1) (inc2 ci 1))
     (-> (PC_r7 8 8) ebus1 (inc2 i))
     (function (inc) inc2)
     (-> (inc2 o) inc2o)
state44
     (memory (busreadcontinue) M_r2)
     (-> edb di)
     (-> inc1o ebus1 (PC_r7 8))
     (-> inc2o ebus1 (PC_r7 8 8))
state45
     (memory (busreadbegin) M_r2)
     (-> di ebus1 (ao 8))
     (-> 0 ebus1 (ao 8 8))
     (-> (ao 8) (eab 8))
     (-> (ao 8 8) (eab 8 8))
state46
     (memory (busreadcontinue) M_r2)
     (-> edb di)
state50
     (-> di ebus1 tn3)
     (jmap (ipla2) ire0)
```

Figure 7-18
Final Compacted Control Sequence for *option1* after Scheduling and Functional Unit Binding

Option2 is also not eliminated because its minimum functional unit cost is found to be 10 cost units and its mean function concurrency is 1.0. *Option3* is eliminated in the first pass of code evaluation since it requires at least 2 8-bit arithmetic-logic units (see *state10* in Figure 7-17) while the constraints allow at most one arithmetic-logic unit. Since the maximum mean function concurrency is 2.0 and the mean function concurrency of *option2* is less than the maximum mean function concurrency, *option2* is eliminated in the second pass of code evaluation. Therefore, *option1* is the best code option.

After data forwarding *option1*, the code scheduler tries to design the fastest compacted control sequence. The functional unit assignment algorithm then assigns the structural functional unit *inc1* to the symbolic functional unit *inctn1* since they both belong to the processing section *al*. The structural functional unit *inc2* is assigned to the symbolic functional unit *inctn0* since they both belong to the processing section *ah*. *Option1* is then scheduled again to get a more accurate compacted control sequence. The final schedule is shown in Figure 7-18. The speed and cost of the compacted control sequence are measured and found to be 10 clock cycles[3] and 12 cost units respectively. The third pass of code evaluation is trivial since there is only one feasible code option. *Option1* is therefore the most desirable code option.

7.5.2. Observations About Code Selection

Owing to complex interactions between the various subproblems of code selection, some form of phase coupling is desirable. SUGAR tries to reduce interactions by using novel methods, and delaying decisions as long as possible so that the design is not unnecessarily constrained. The three pass method used to evaluate code options appears to produce good results for the microprocessor design style but may not apply to other design styles because the method requires designs with low parallelism. Experimental results are presented in Chapter 8. The functional unit assignment algorithm only uses information local to a code option. The lack of global information is not a drawback if the interconnections are fixed as in designs synthesized using fixed bus models but could be a drawback if the busses are synthesized bottom-up because the interconnections in such designs are affected by the functional unit assignment. However, the drawback does not appear serious because the low parallelism of the microprocessors probably limits the number of feasible symbolic functional unit to structural functional unit assignments.

7.6. REGISTER AND BUS ASSIGNMENT

Hardware assignment is the assignment of symbolic units to structural units. A *symbolic unit* is a data structure that is used by SUGAR to record information needed for assignment. Examples of information recorded in a symbolic unit are its size and restrictions on

---

[3] A 4-phase clocking scheme is assumed in which the duration of each control state is 2 clock cycles. Since there are 5 control states in the schedule, the schedule requires 10 clock cycles.

the types of structural units to which it may be assigned. Symbolic units are created by SUGAR before the assignment phases.

Registers are storage units used to hold data. Symbolic registers are of two types: symbolic temporary registers and symbolic architectural registers. *Symbolic temporary registers* are symbolic registers used to describe structural registers needed to hold the values of temporary variables and intermediate expressions in the given behavior. *Symbolic architectural registers* are symbolic registers used to describe structural registers visible to the microprocessor's assembly language programmer. Register assignment is the assignment of symbolic registers to structural registers. Busses are interconnection units used to transmit data. *Busses* and other types of interconnection units are described in greater detail in Section 7.6.3. Bus assignment is the assignment of symbolic busses to structural busses. Bus assignment is done after register assignment. This section describes the register and bus assignment phases of SUGAR.

Assignment may be viewed as a packing problem where the goal is to pack the symbolic units into the fewest structural units. The packing problem in SUGAR is formulated in terms of coloring an *interference graph*. Graph coloring attempts to use the minimum number of colors when assigning a color for each node of a graph such that nodes with the same color have no edges between them. In contrast to the clique partitioning used by Busser which operates on a compatibility graph and looks for the minimum number of clusters such that each cluster is made up of complete graphs, graph coloring operates on an interference graph which is the complement of the compatibility graph, and looks for the minimum number of clusters such that each cluster has no edges between its nodes. To use graph coloring in mapping behavior onto structural units, the vertices of the graph represent the symbolic units. An edge connects two vertices if the two symbolic units interfere, that is, they cannot share the same structural unit.

7.6.1. Graph Coloring Algorithm

Given a reasonably large graph, say with 20 vertices or more, it is difficult to determine the best coloring for it. SUGAR uses the Recursive Largest First Algorithm (RLF) algorithm of Leighton [Leighton79] since it appears to be the best heuristic algorithm for coloring graphs. The algorithm is given in Figure 7-19.

```
1.  i = 0
2.  While (vertices to color) {
3.      i++
4.      Color the vertex that has the maximal degree (maximum
                    number of edges) with color i
5.      U1 = set of uncolored vertices not adjacent to any vertex of
                    color i
6.      While (U1 is not empty) {
7.          U2 = set of uncolored vertices adjacent to at least 1 vertex
                    colored i
8.          Find the vertex (Vi) from U1 that is connected to the most
                    vertices in U2
9.          If (more than 1 such vertices Vi from step 8.) {
10.         Choose the vertex Vi that has the fewest connections in U1
            }
11.     Color Vi with color i
12.     Recalculate U1
        }
    }
```

Figure 7-19. Recursive Largest First Algorithm for Coloring Graphs

The RLF algorithm requires $O(n^3)$ time and $O(n^2)$ space to color an n vertex graph. Leighton has shown that for sparse graphs the algorithm requires only $O(n^2)$ time.

7.6.2. Register Assignment

The register assignment phase builds a symbolic register interference graph, colors the graph and then binds the assigned registers to the design. Two symbolic registers are said to *interfere* if one is defined at the point where the other is live and the two registers do not have the same value. Since the packing method used requires registers of uniform size, symbolic registers are partitioned into disjoint sets such that all members of a set have the same size.

The fixed bus models used in SUGAR assume that the execution unit is partitioned into processing sections that contain local functional units and local registers. The communication between the processing sections is organized in such a way that maximum parallelism can be achieved only if the functional units access registers local to their section. Therefore, whenever a fixed bus model is used, symbolic registers are partitioned into disjoint sets such that all members of each set belong to the same processing section. This ensures that a symbolic register will be compatible only with other symbolic registers belonging to the same processing section. In addition, it ensures that the control sequences will not be invalidated by the register assignment since the sequences were synthesized for a particular interconnection structure.

If a fixed bus model is not used, i.e., the busses are synthesized bottom-up, symbolic registers belonging to different sections can be assigned to the same structural register since the interconnection structure for such designs is synthesized after register assignment. In such instances the restrictions described in the preceding paragraph do not apply.

7.6.3. Bus Assignment

This section describes the assignment of symbolic busses to structural busses. SUGAR can synthesize busses using a top-down or bottom-up method. Both synthesis methods assign symbolic busses to structural busses by building a symbolic bus interference graph and coloring it.

The top-down synthesis method fixes the global interconnections early in the design by selecting a bus model appropriate for the behavior. Each bus model supported by SUGAR has two structural busses. The bus assignment phase assigns each symbolic bus to one of the two structural busses. The top-down bus synthesis method also uses the bus assignment method to analyze bus conflicts during code scheduling. Since the top-down bus synthesis method uses bus models with fixed global interconnections, the code scheduler has to ensure that control sequences it designs are feasible for the given interconnections. The conflict analysis method is discussed later in this section.

In contrast to the top-down synthesis method, the bottom-up synthesis method does not fix the global interconnections or restrict the number of structural busses. Instead, symbolic busses are packed into the fewest structural busses.

Bus Interference Graph. The bus interference graph consists of vertices which represent symbolic busses. Two symbolic busses *interfere* if they are used in parallel and are driven by different sources. This definition needs to be relaxed for top-down bus synthesis to ensure that the special bus topologies of the bus models are used efficiently.

Recall that in top-down bus synthesis with fixed bus models the execution unit is partitioned into processing sections that contain local functional units and local registers. For instance, consider the Motorola MC68000 bus model shown in Figure 7-20. The execution unit in this model consists of 3 processing sections: an address high/data high section *(ah)*, an address low section *(al)* and a data low section *(dl)*. The 2 structural busses *ebusa* and *ebusb* are partitioned by independently controllable bus couplers *C1*, *C2*, *C3* and *C4* into 6 busses *ebusa-ah*, *ebusa-al*, *ebusa-dl*, *ebusb-ah*, *ebusb-al* and *ebusb-dl*. Concurrent bus transfers are possible on different sections of the same structural busses if the sections are isolated by disabling the bus couplers between them. Up to 6 parallel bus transfers are possible if all 4 bus couplers are disabled while only 2 parallel bus transfers are possible if all 4 bus couplers are enabled.

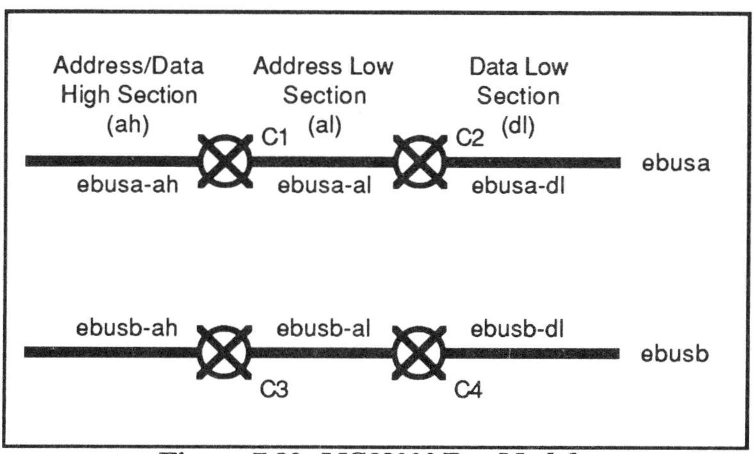

Figure 7-20. MC68000 Bus Model

Since two bus transfers on a structural bus isolated by a disabled bus coupler do not conflict, two symbolic busses in the MC68000 bus model are said to *interfere* if all the following conditions are satisfied:

- They are used in parallel.

- They are driven by different sources.
- They move data through the same processing sections.

For instance, let *b1* and *b2* be two symbolic busses used in parallel and driven by different sources. If *b1* moves data from section ah to section *al* and *b2* moves data from section *al* to *dl*, then *b1* and *b2* interfere because they both move data through section *al*. If instead, *b2* moves data only within section *dl*, then *b1* and *b2* would not interfere.

The IBM Micro 370 bus model shown in Figure 7-21 is the only other bus model supported by SUGAR. This bus model has 2 processing sections: an address section *(a)* and a data section *(d)*. The 2 structural busses *ebusa* and *ebusb* are partitioned by bus couplers *C1* and *C2* into 4 busses *ebusa-a*, *ebusa-d*, *ebusb-a* and *ebusb-d*. Up to 4 parallel bus transfers are possible if both bus couplers are disabled. The definition of bus interference for the IBM Micro 370 bus model is same as that for the MC68000 bus model.

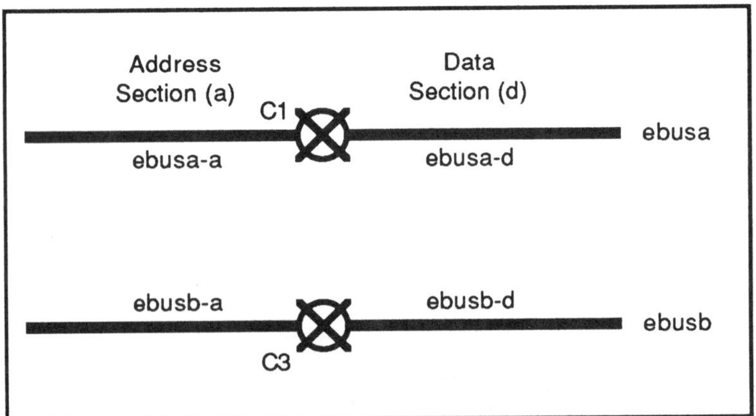

Figure 7-21. IBM Micro 370 Bus Model

The interference graph for top-down bus synthesis (MC68000 and Micro 370 bus models) is built for each control state as follows:

1. Determine for each symbolic bus what execution unit processing sections are used to move data.

2. For each pair of parallel bus transfers represented by symbolic busses, add an edge to the interference graph if the busses are driven by different sources over the same processing sections.

For bottom-up bus synthesis, a good bus assignment method should ensure that symbolic busses which do not share a source or a destination are never assigned to the same structural bus. For bottom-up bus synthesis, 2 symbolic busses are said to *interfere* if both the following conditions are satisfied:

- They are used in parallel.
- They are driven by different sources.

The interference graph for bottom-up bus synthesis is built by adding an edge to the interference graph for each pair of symbolic busses which satisfy the above conditions. Unlike top-down bus synthesis, only one interference graph is built since all bus transfers in the design are considered. As in Busser, symbolic busses may be combined onto a common structural bus only if the symbolic busses have no sources or destinations in common. This is likely to produce a large number of busses. A global bussing phase similar to that used by Busser has the potential to address this problem by using the high level address and data partitions as global endpoints, allowing busses to be combined.

Conflict Analysis in Top-down Bus Synthesis. The top-down bus synthesis method uses 2-bus models with fixed topologies. Since the fixed topologies limit the parallelism of the system, not all control sequences synthesized by the code scheduler are feasible. The inner loop of the code scheduler is a hardware conflict analysis algorithm that determines if two sets of microoperations conflict when packed into a control state. For top-down bus synthesis, interconnection conflicts are determined as follows:

1. A dummy control state containing the two sets of microoperations is built.

2. A bus interference graph is built as explained earlier in this section.

3. The microoperations conflict if more than two colors are required to color the graph since this means that more than two structural busses will be needed to execute the microoperations in parallel.

7.6.4. Observations about Register and Bus Assignment

The interference graphs for register and bus assignment are usually sparse compared to the corresponding compatibility graphs. Bhasker and Samad conjecture [Bhasker87] that coloring algorithms outperform clique partitioning algorithms for sparse graphs.

The top-down method normally leads to better layouts since the interconnections are fixed before the design is synthesized. The gain in layout quality sometimes comes with some loss of performance since the fixed bus models limit the available parallelism.

7.7. PHASE STRUCTURE OF SUGAR

The preceding sections described the important phases in the multi-phase method used by SUGAR to synthesize microprocessors. Descriptions of phases which were not described can be found in [Rajan88]. This section draws on the earlier descriptions to show how the phases are organized and how they communicate with each other.

Each phase solves one or more subproblems in synthesis. The phases communicate with each other through internal behavioral descriptions which are either TCOL syntax trees or data flow graphs. Each phase transforms, annotates or translates its input. LISP property lists, the primary communication mechanism between phases, are used to record descriptions of design decisions and design analyses. Information recorded by earlier phases is kept as long as it is needed by later phases.

Figure 7-22 shows how the phases in SUGAR are organized. The figure identifies the types of knowledge representations, the types of behavioral representations passed between phases, and the major interactions between phases. While most phases use an algorithmic knowledge representation, a few, like code generation, truncation transformation, expression context analysis and constant folding, use a combination of rule-based and algorithmic knowledge representations. Most phases operate on TCOL syntax trees, but a few operate on data flow graphs.

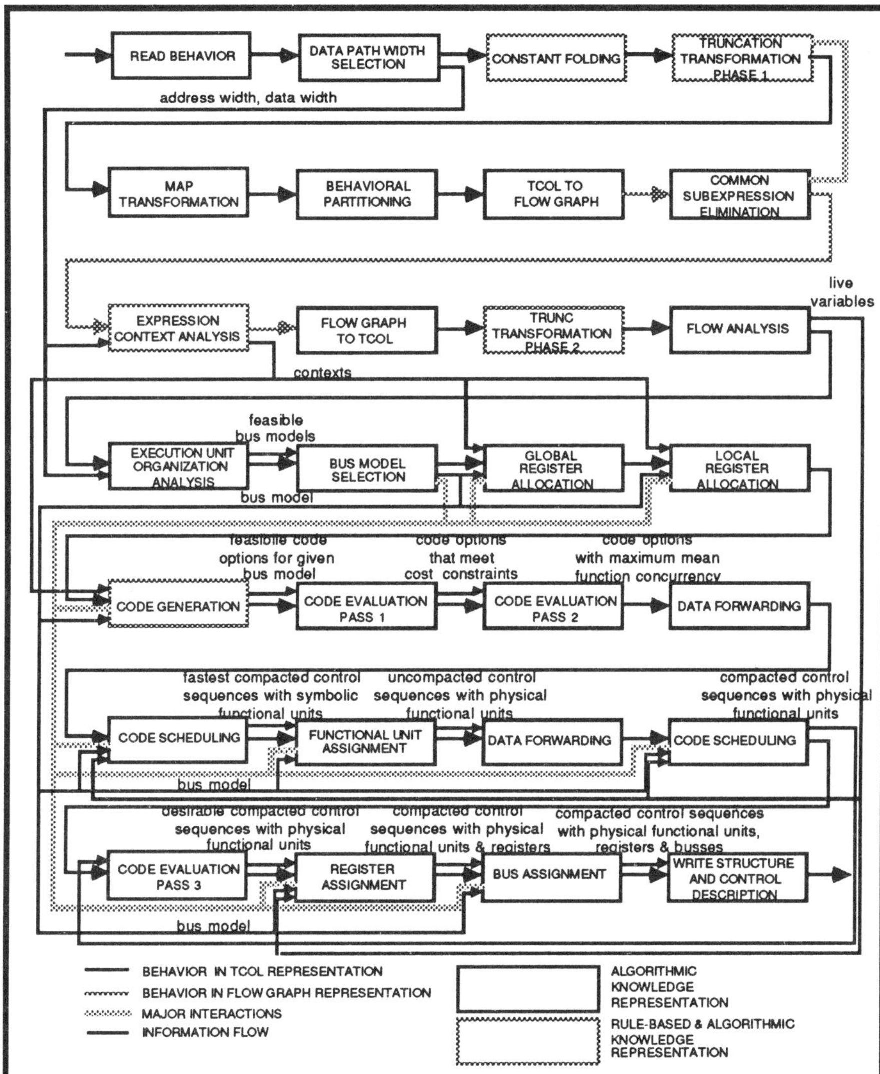

Figure 7-22. Organization of Phases in SUGAR

A delayed binding strategy is the primary method used to resolve interactions between phases. Decisions are delayed until enough information is available to make them. Some phases are repeated so that conservative design decisions made in an earlier pass can be refined in a later pass based on design decisions made by other intervening phases. For instance, the truncation transformation phase is executed twice to resolve its interaction with the common subexpression elimination phase.

A three pass code selection strategy using the delayed binding principle is used to select the desirable code options from the feasible code options generated by the code generation phase. The code evaluation pass 1 phase and the code evaluation pass 2 phase use simple analyses to eliminate undesirable code option generated by the code generation phase. The code scheduler then designs the fastest control sequences for the given bus model using inaccurate information about functional unit bindings since functional units have not been assigned at this stage. The sequences generated are used by the functional unit assignment phase to guide the assignment of functional units. In a later phase, the code scheduler uses the results of the functional unit assignment to synthesize more accurate control sequences.

The delayed binding method has the advantage that the functional unit assignment phase and the code scheduling phase take into account the global interconnections described by the bus models. Details about structural registers and structural busses are added to the design by the register assignment phase and the bus assignment phase respectively only after the desirable control sequences have been synthesized.

While SUGAR's complex multi-phase strategy has proved to be useful in resolving some difficult interactions in microprocessor synthesis, additional work is necessary to resolve interactions which SUGAR does not address. Of these, the most serious interactions are those between the transformation phases and the synthesis phases.

SUGAR is written in FRANZ LISP, a dialect of LISP, and consists of more than 35 phases and over 20,000 lines of code. Analysis of the performance of the program is difficult since it contains many algorithms and many rules.

7.8. SUMMARY

SUGAR uses a style-specific synthesis method for the microprocessor design style. The method synthesizes even large microprocessors like the MC68000 rapidly in only 80 CPU minutes on a VAX 8800. Although definite conclusions in favor of the style-specific method will require more experimentation, the results presented in the next chapter suggest that the microprocessor style-specific method used by SUGAR synthesizes better designs than the general methods used by other synthesis programs.

SUGAR contains knowledge about microprocessor interconnections represented as a set of bus models that can be used to synthesize busses top-down. Two bus models, namely, the IBM Micro 370 and the Motorola MC68000 bus models were implemented. Experimental comparisons of top-down synthesis of busses using fixed bus models with bottom-up synthesis of busses suggest that top-down synthesis of busses with fixed bus models results in designs with better interconnections.

SUGAR uses a delayed binding method to resolve interactions between problems in synthesis. The method is used developed to solve the particularly knotty circular dependency between allocation and scheduling.

SUGAR uses a combination of algorithmic and rule-based knowledge representations. Several problems such as behavioral transformations and register assignment are solved using algorithms since efficient algorithms exist for these problems. Other problems such as code generation are solved with a rule-based approach since the design space for these problems is large and the underlying complexity of the problems is high.

The structure of SUGAR shows that an automatic synthesis program for the synthesis of microprocessors is a complex program that requires careful thought and planning. SUGAR's phase organization is an attempt to solve the complexity problem.

Results of using SUGAR to synthesize two microprocessors are given in the next chapter, and are compared with results from other synthesis programs and with commercial designs.

8. Synthesis Results

Designs for five architectures are presented in this chapter to illustrate the range of capabilities of the synthesis tools in the Workbench. The designs, summarized in Table 8-1, are listed in order of increasing complexity. The first 4 architectures, the Elliptical Filter, the Kalman Filter, the BTL310, and the MCS6502 were synthesized using various design exploration options in the general synthesis path. The last two architectures, the MCS6502 and the Motorola 68000 are both microprocessors. The MCS6502 was synthesized through both the general and style specific paths. The 68000 is too big for the general path, but a portion of its capabilities was synthesized by SUGAR.

The simplest architecture presented here is a small elliptical filter that can be described with a single vtbody. All of the general synthesis path tools were used to synthesize various designs for this architecture. The designs demonstrate the effects of behavioral transformations and architectural partitioning on the design. The second architecture is also a filter, the Kalman Filter. Several designs of this architecture show how the EMUCS cost options can be used to explore the design space. A DSP machine, the BTL310, is also presented in this chapter. This is an important design because it can be compared with an actual commercial design. The two microprocessors presented in this chapter are the MCS6502 and the Motorola 68000. The MCS6502 is a relatively

Table 8-1. Designs Presented in this Chapter

Design	Behav'l Trans'ns	APARTY	CSTEP	EMUCS	Busser	SUGAR
Elliptical Filter	X	X	X	X	X	
Kalman Filter		X	X	X	X	
BTL310		X	X	X	X	
MCS6502	X	X	X	X	X	X
68000						X

small microprocessor that can be synthesized through each of the Workbench's major synthesis paths, allowing the resultant designs to be compared both with each other and with the commercial design. The 68000 is a relatively large architecture that was synthesized by SUGAR and is compared with the commercial design.

There are a number of criteria that can be used for evaluating a design. Optimally, silicon area and actual performance measures would be presented for each design. However, these values are usually unavailable for the Workbench designs because the link between the Workbench and lower level synthesis tools has not been completed. The Workbench has recently been integrated with the LASSIE layout tool, which is currently under development [Trick89]. LASSIE has been used to lay out the Elliptical Filter, so actual area measures are available for this machine. LASSIE is not yet able to layout any of the other Workbench designs presented in this chapter, so other evaluation criteria are used to measure the cost in terms of the area and the performance of a design. These measures allow for comparisons between designs.

Several metrics are used to measure the *area-cost* of a design. The area is measured indirectly in terms of how many components are used, the bits of hardware used and the size of the control unit. The metrics are as follows:

1. *Control Unit:* The size of the control unit can be estimated by the number of control points required to control the components in the

microprocessor (this determines the number of bits of microcode needed) and the number of control states (microwords, or csteps) there are in the design. Another important measure is the number of unique states.

2. *Functional Units:* Total number of bits of functional units, where a functional unit implies combinational logic to implement a function, such as an adder.

3. *Wiring Units:* Total number of bits of wiring units, where a wiring unit implies wires that implement a function such as a concatenation of values.

4. *Registers:* Total number of register bits.

5. *Interconnections:* Multiplexors and busses represent actual combinational logic, and imply the degree of interconnection complexity for a design. The complexity can be broken down further by considering the numbers of inputs and outputs of the multiplexors and busses. Since bus inputs require tri-state devices, there is a cost associated with them, while bus outputs may require drivers, which also add to the area of a design.

Caveat: EMUCS's designs have some error due to mismatched bitwidths, so EMUCS bit counts presented in this chapter are estimates.

The *performance* of a design can be evaluated by considering the control step schedule. The number of csteps is the most basic measure of performance, and it can be applied to any design. One specific performance measure used for the microprocessor designs is the *average execution time.* The average execution time is the average number of control states executed for each instruction and is computed on the assumption that all execution paths in a routine are equally likely.

The above measures are used to evaluate the area and performance of the various designs for the 5 architectures presented in this chapter.

8.1. FIFTH ORDER DIGITAL ELLIPTIC WAVE FILTER

This example, a fifth order digital elliptic wave filter from [Dewilde85], was one of the examples from the ACM / IEEE 1988

Workshop on High-Level Synthesis [Borriello88]. The VT for the Elliptical Filter is shown in Figure 8-1.

Several design permutations for the Elliptical Filter are presented in this section. The permutations are listed in Table 8-2 as design A-E. The design options used in the synthesis of each design are listed vertically. These designs demonstrate the result of using behavioral transformations to create separate processes and the effect of architectural partitioning on the design. All of the designs were run through CSTEP and EMUCS to complete the RT designs which are compared at the end of this section. In order to do a thorough comparison of the designs, two of the final data path designs were layed out using the LASSIE tool to obtain actual area figures for these designs. The designs will all be discussed according to the tools used for the synthesis of each one.

Table 8-2. Summary of Synthesis Options for Elliptical Filter Designs

Tool	Design				
	A	B1, B2, B3	C	D	E
Behavioral Transforms	none	2 horizontal processes	2 vertical processes	none	none
Architectural Partitioning	none	none	none	none	2 data partitions
CSTEP	ASAP	ASAP	ASAP	max of 2 adders, 2 multipliers	max of 1 adder, 1 multiplier per partition
EMUCS	default costs	default costs	default costs	default costs	default costs
Busser	not run	not run	not run	default	default
LASSIE	not run	not run	not run	default	default

8.1.1. Behavioral Transformations

Behavioral transformations were run on 4 of the Elliptical Filter designs to create multiple process designs. Ultimately, the RT designs that result from these transformations are compared with the single process design, A. This section describes the effects of splitting the VT horizontally, encapsulating the operators in the gray rectangle in Figure 8-1 into a separate process to produce design B2. Designs B1 and B3 are produced by similar horizontal cuts just above operator x3 and x26, respectively. This section also describes the effects of splitting the VT vertically, encapsulating the operators in the dark gray oval of Figure 8-1 into a separate process to produce design C. Further details on these process divisions can be found in [Walker88, Chapter 6]. Both the horizontal and vertical divisions are described below.

Horizontal Two-Process Filter. The Elliptical Filter VT can be cut horizontally to produce a two-process design. One possible horizontal cut is shown by the gray rectangle in Figure 8-1. If the operators in the gray rectangle of Figure 8-1 are to be transformed into a second process, the first step is to encapsulate those operators into a new vtbody using Vtbody Formation. Once the new vtbody has been formed, it can be converted into a separate process using Process Creation. Other process divisions can also be explored; this division was chosen by the designer to split the VT horizontally into two processes with roughly the same number of operators and control steps. This design, cut at operator x16, is represented as design B2 in Table 8-2. Two other horizontal cuts, at x3 and x26, are also presented in this section, and are referred to as B1 and B3, respectively.

Vertical Two-Process Filter. Alternately the Elliptical Filter's VT can be split vertically to form another two-process design. In Figure 8-1, the dark gray oval becomes the second process. The design that is the result of this process division is referred to as design C in Table 8-2.

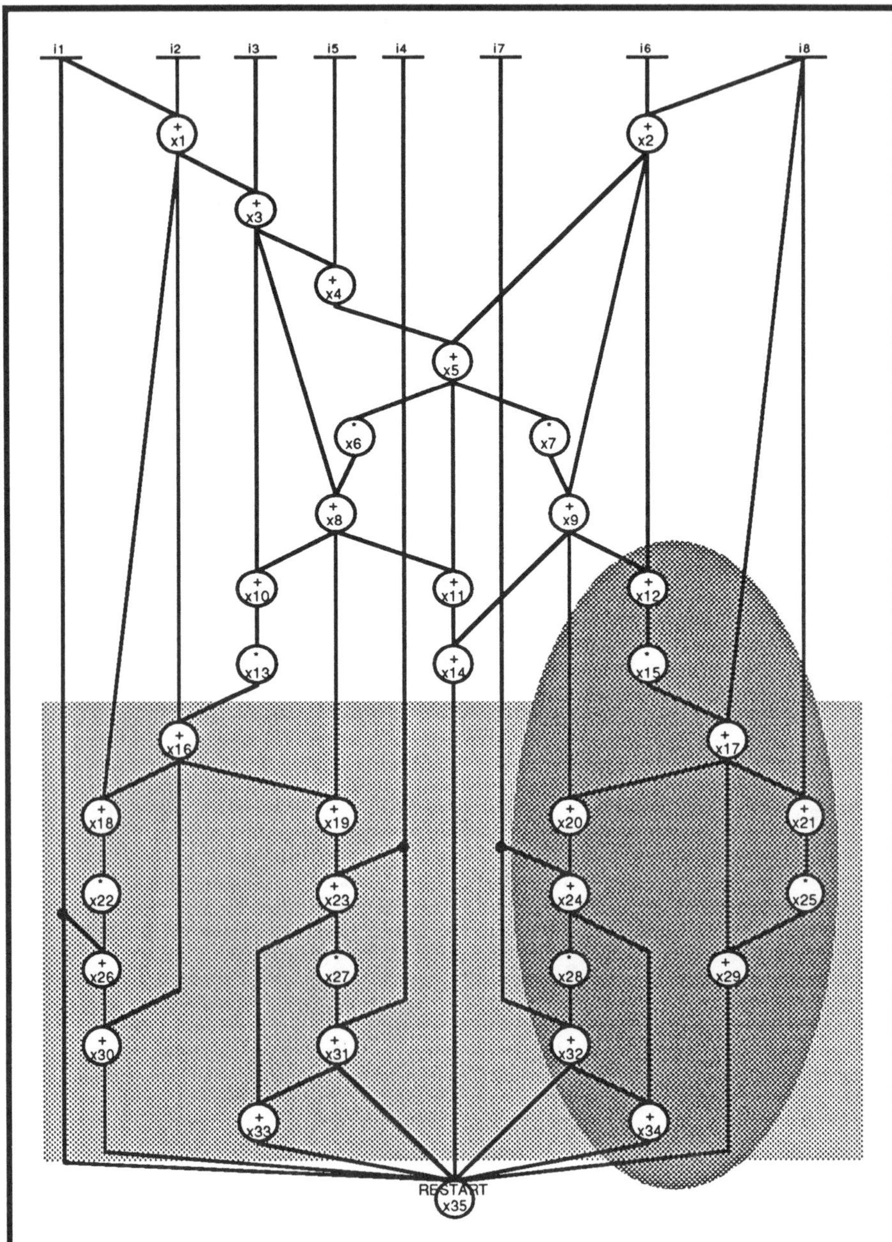

Figure 8-1. Filter VT Before Process Creation

8.1.2. Architectural Partitioning

While designs B1, B2, B3, and C were partitioned by hand into multi-process descriptions, and design D was not partitioned at all, design E was partitioned by APARTY to produce a single-process partitioned design. Four stages of data clustering were used to create the partition shown in Figure 8-2. In the figure, the partitions are delineated by the shaded lines. There are several interesting aspects to this partitioning of the Elliptical Filter.

Unlike the partitions chosen by hand in the previous section for designs B and C, the partitions chosen by APARTY and shown in Figure 8-2 were not chosen with high level concurrency in mind. However, the low number of data dependencies between the two partitions make the creation of processes from the partitions an interesting prospect. Although the current transformations cannot handle the multiple entry points that would be needed to create these processes, such a transformation could be defined.

There are PLUS and MULT operations in each partition. Thus, in order to implement the partitions, the design requires at least 2 adders and 2 multipliers (one of each per partition). This information is passed on to both CSTEP and EMUCS, and their use of the information is described in the next sections.

There are only three data values passed between the two partitions (one of those values which is produced by the plus operator in cstep 4 in the left partition is consumed by two operators in the right partition). This low degree of interconnect indicates that the goal of the data clustering was met, and allows for a good data path allocation and good routing of the design. This design will be compared directly with design D, which was synthesized without partitioning.

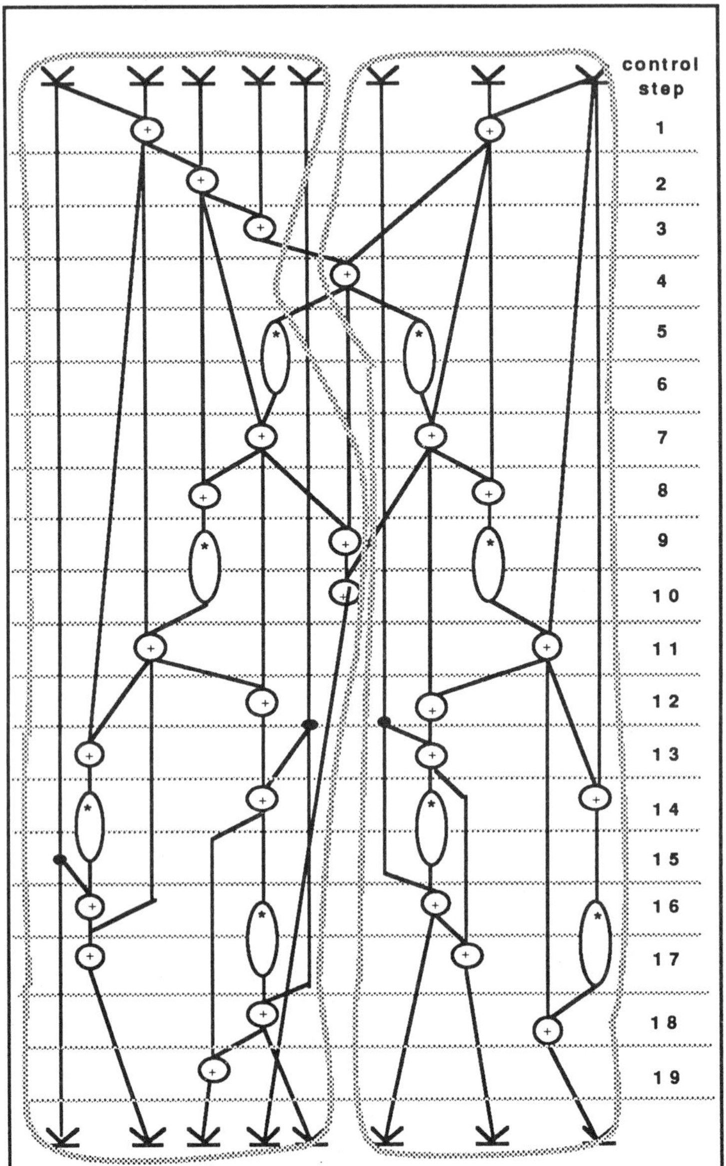

Figure 8-2. Partition of Elliptical Filter Description

8.1.3. Scheduling

Different scheduling options were used to allow specific comparisons among the designs. All designs use the assumption that a MULT operation executes in 2 csteps while a PLUS operation requires only 1 cstep. As indicated in Table 8-2, designs A, B, and C were all scheduled ASAP. Since no consideration was given to resource usage for these designs, they have short schedules that require a lot of hardware. The remaining 2 designs used limited hardware resulting in longer schedules.

The minimum hardware is used for the partitioned design. So that the final designs can be compared, the unpartitioned design (D) uses comparable hardware constraints. Thus, both D and E require 2 adders and 2 multipliers, but design E is further constrained in that it allows 1 adder and 1 multiplier in each of its 2 partitions, so that an operator can only be scheduled if the appropriate functional unit in its partition is free. A 19 cstep schedule was chosen for each, but since the unpartitioned design does not use a minimum hardware assumption, thus requiring constraints from the designer, it is clear that partitioning provides valuable information for the scheduling phase. The partitioned schedule is shown at the right of Figure 8-2.

Some other scheduling results for the Elliptical Filter are listed in Table 5-1.

The CSTEP schedule lengths and the implied microcode sizes for all designs are shown in Table 8-3. For each design, the table lists:

bits of μcode for control flow– the total number of bits of microcode required to manage control flow in each process, based on the total number of states produced by CSTEP and an assumed 4 bits to specify the microoperation

bits of μcode for data path – the total number of bits of microcode required to manage the data path in each process

change – the percent change in the total number of bits of microcode needed to control the data path between the single-process design and the two-process design *total μcode change* – the total percent change in the number of bits of microcode between the single-process design and the two-process design

Table 8-3

Effect of Process Creation and Partitioning on Filter's Microcode

design	description	μwords (CSTEP)	bits of μcode for control flow	change	bits of μcode for data path	change	total μcode change
A	single process, unlimited hardware	17	153		680		
B1	horizontal split at x3	1 16	5 128	-13.1%	8 720	7.1%	3.4%
B2	horizontal split at x16	10 7	80 49	-15.7%	240 245	-28.7%	-26.3%
B3	horizontal split at x26	14 4	112 24	-11.1%	560 84	-5.3%	-6.4%
C	vertical split	17 10	153 80	52.3%	544 150	2.1%	11.3%
D	single process, limited hardware	19	171	11.8%	589	-13.4%	-8.8%
E	single process, partitioned	19	171	11.8%	589	13.4%	-8.8%

Effect of Process Creation on Microcode. This table shows several effects of Process Creation on the design's microcode. For the horizontally split designs (B1, B2, B3), the table shows that only design B3 requires an extra control step for its ASAP schedule. There is an 11 to 16 percent decrease in the amount of microcode required for control, and between a 7 percent increase and a 29 percent decrease in the amount of microcode required to control the data path due to splitting the architecture into two processes. Finally, the table shows a total microcode change of between a 3 percent increase and a 26 percent decrease for the horizontal splits. For the vertically split description (C), the two processes can be run concurrently, so the total time for execution is still 17 csteps for the ASAP schedule. Since the first process alone requires the same number of control steps as the single-process design, the microcode required for control increases. The total number of bits required to control the data path decreases, but not enough to offset the additional amount necessary for control, so the overall effect is an increase in the amount of microcode.

As predicted in Section 3.4.3, the largest decrease in the amount of microcode required for control occurs when the two stages are split to contain an equal number of control steps. In this example, the division closest to meeting this criterion is the split before x16 (Design B2), where the two processes contain 10 and 7 control steps, respectively; this split produces a 15.7% decrease in the microcode required for control. As the split is moved away from this division, as shown by designs B1 and B3, there is a smaller improvement in the size of the microcode.

The number of bits of microcode to control the data path is dependent on the data paths allocated by EMUCS for each design. The data paths are described in the next section, but the effect on the microcode is described here. Intuitively, more hardware modules require more bits in the microword to control them. However, there may be a decrease in the microcode even when the total amount of hardware in the data path increases. For example, before Process Creation, the filter design has 17 microwords, each requiring 29 multiplexor control bits and 11 register control bits, for a total of 40 bits to control the data path. For the split before x16, the first process has 10 microwords, each requiring only 14 + 10 = 24 bits to control the data path, and the second process has 7 microwords, each requiring only 26 + 9 = 35 bits to control the data path. Even though the total amount of hardware has increased (see Table 8-4), the microcode decreases due to the decrease in the width of the microwords.

Effect of Partitions on Microcode. Since both design D and design E have longer schedules, they require more bits of microcode for control. However, the longer schedules are a result of limited hardware resources, which results in fewer bits of microcode to control the data path. Both of these results are reflected in Table 8-3. Microcode size is the same for designs D and E because they have the same schedule length and essentially the same hardware; design E requires fewer multiplexor inputs but the same number of multiplexor control bits as does design D.

As mentioned previously, forming processes from the APARTY produced partitions is an interesting consideration. If the necessary transformation were available to do this, the overall controller size for the new two-process design would likely be larger than for the single-process design because a separate controller would be required for each processor. However, each of these controllers would be smaller than the controller needed to implement this data-flow as a whole, because they would each be controlling fewer modules. The schedule that is indicated by the control steps listed at the right of Figure 8-2 could be used to run the processes concurrently. Using this schedule, the processes would not require extra control steps to run in lockstep because the data is already being passed between the partitions at control step boundaries.

Different scheduling options have been presented for the various partitionings of the Elliptical Filter. Designs A, B, and C used ASAP scheduling, where an unlimited number of functional units is assumed. Design E uses the minimum number of functional units for the partitions, while Design D assumes a comparable number of functional units, although this is not the minimum for the unpartitioned design. After scheduling, data path allocation is the next step for each of the designs.

8.1.4. Data Path Allocation

Each of the designs was allocated using default costs. The default costs suggest that the minimum number of functional units and registers be used for each design, so that the differences due to the other design aspects can be compared.

Table 8-4 shows several effects of Process Creation and partitioning on the data path. Area estimates are based on MOSIS 3μm CMOS standard cells; for details see [Walker88]. For the multi-process designs, the last column of the table lists *% of single process design* – the ratio of the area of each process of the two-process designs to the total area of the single process design.

Effect of Process Creation on Data Path. The change in the estimated area as a result of Process Creation is important particularly if the designer's goal is to split a large design across two chips. For example, for Design B2, splitting the VT at x16 and assuming unlimited hardware results in a two-process design with a small first process, but with a second process using the same number of multiplication and addition functional units as the single-process design. Since the multiplication functional units dominate the area estimate, the second process alone is not much smaller than the single-process design. If the designer's goal in using Process Creation was to

Table 8-4. Effect of Process Creation on Filter's Data Path

design	description	values xferred between processes/ partitions	add. funct. units (EMUCS)	mult. funct. units (EMUCS)	registers (EMUCS)	mux input bits (EMUCS)	total area (sq. mm)	% of single process design
A	single process, unlimited hardware		4	4	11	848	1036.7	
B1	horizontal	3	2	0	8	0	2.7	0.3%
	split at x3	2	4	4	10	1024	1039.7	100.3%
B2	horizontal	6	3	2	10	400	519.3	50.1%
	split at x16	3	4	4	9	704	1034.0	99.7%
B3	horizontal	9	4	2	10	832	527.8	50.9%
	split at x26	6	2	2	8	384	517.9	50.0%
C	vertical	3	2	2	10	608	521.9	50.3%
	split	2	2	2	4	304	516.1	49.8%
D	single process, limited hardware		2	2	11	560	521.3	-49.7%
E	single process,	1	1	1	7			
	partitioned	2	1	1	5	544	521.0	-49.7%

split a large design across two smaller chips, these results are unacceptable. One possibility would be to limit the number of functional units in the design, incurring a penalty in the schedule length. In contrast, splitting the VT at x26 as was done for design B3 results in two processes, each of which is approximately half the size of the single-process design, and possibly able to fit onto the chip, although this design's ASAP schedule requires one more control step than either of the other two horizontal splits.

For the vertically split design, the total number of addition functional units and multiplication functional units in the two-process design matches that of the single-process design, meaning that no additional functional units were added due to Process Creation. The number of registers and multiplexor input bits increased slightly due to the communication overhead. Since the functional units dominate the estimated area, the resulting two-process design requires slightly more area than the single-process design. Depending on the layout, this could result in smaller path lengths, a faster clock, and better performance.

Effect of Partitions on Data Path. As noted earlier, the resource constraints on designs D and E lead to smaller designs. This is reflected in the comparison of the total area for these designs with the single process, unlimited hardware design. Since the total number of functional units and register was the same for designs D and E, the differences between the two designs result from the mapping of the data-flow operators and values onto the data path hardware. For a minimum hardware data path using the given schedule, partitioning fully determines the binding of functional units, and directs the binding of registers according to the partitions. In contrast, when designing without the use of partitioning information, there are a multitude of possible binding options to explore. The difference in the bindings to these hardware units is reflected in the use of multiplexors. The partitioned design requires slightly fewer multiplexors with fewer inputs. This demonstrates that partitioning can provide good guidelines for data path design. By reducing the interconnect between partitions, a good binding for the data path was obtained and the overall interconnection complexity was reduced. For more details, see [Lagnese89a, Lagnese89b]

If the partitioned design were to be implemented as separate processes, the same number of functional units would be needed as is used for the single process partitioned design.

The number of values transferred between the two stages also varies for each design. Unfortunately, the 2-process design with the smallest area (Design B2, split before x26) also has the largest number of values transferred. If each process is implemented on a separate chip, these values will have to be transferred between chips, and may require additional driver circuitry and more pins on the chip. The vertical example divides the two processes so as to greatly reduce this data transfer, and the partitioned design requires only 3 transfers of different data values between partitions (although 1 of the values is consumed twice).

8.1.5. Bus Choosing

After the completion of the EMUCS data path allocation, Busser may be used to select busses for the design. The object of using a bussed strategy for interconnect is to simplify routing. Two of the designs, D and E, were carried through to layout and routing, so only these two were run through Busser.

By adding busses appropriately, as described in Chapter EMUCS, the number of connections between the implementation of the partitions can be reduced to 1. Thus, by using data clustering, APARTY achieved its goal of reducing data dependencies between the partitions.

Busser chose busses for both the partitioned and the unpartitioned designs. For the partitioned design, Busser was able to choose busses for the inter-partition connections only, resulting in a single bus which is the sole connection between the partitions. Without the partitioning information to guide the bus selection, Busser chose 7 busses for the unpartitioned design.

The data path for the unpartitioned design is shown in Figure 8-3, while the data path for the partitioned design is shown in 8-4. The data path partitions can be seen clearly in Figure 8-4, where the left side of the data path implements the operators in the left side of the data-flow partitioning scheme. The addition of busses to the design reduces the number of connections between the two data path partitions to a single bus (shown vertically in the figure).

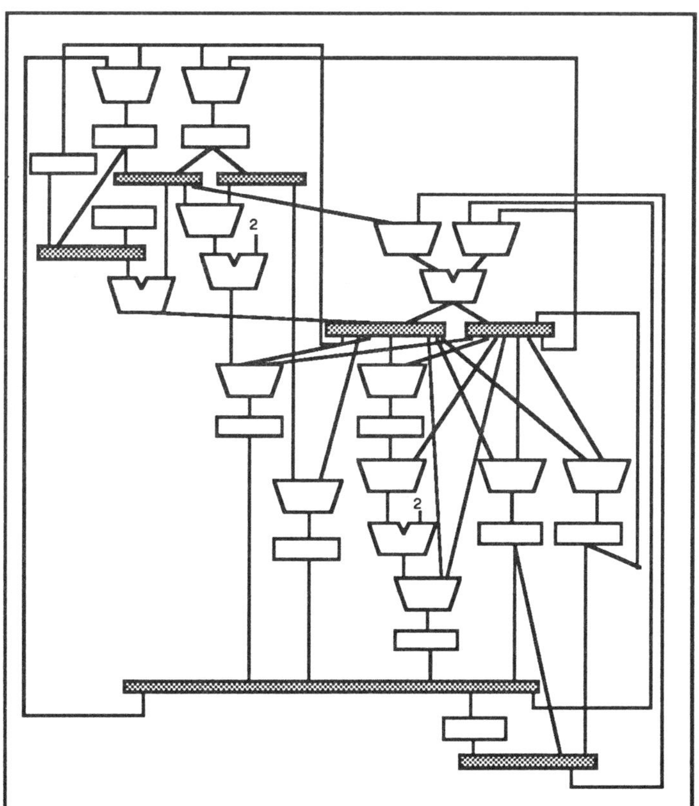

Figure 8-3. Data Path for Unpartitioned Elliptical Filter Design

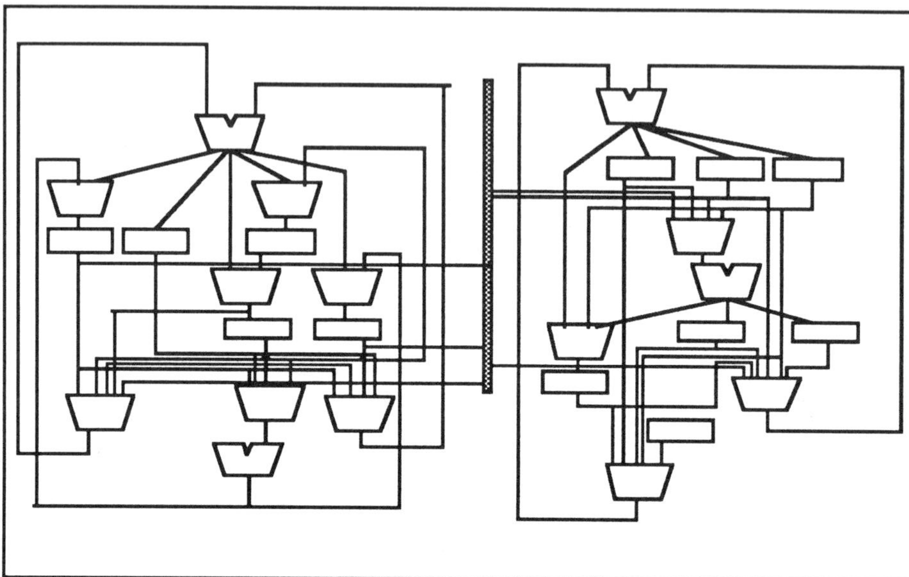

Figure 8-4. Data Path for Partitioned Elliptical Filter Design

8.1.6. Layout

One measure of the complexity of the interconnection scheme is the number of tracks required to route the design, which also affects the overall size of the layout. Both design D and design E were layed out using LASSIE, a layout synthesis tool developed at Carnegie Mellon University. LASSIE used 8 tracks to route the partitioned design, and 10 for the unpartitioned design. A further reduction of 1 routing track appears to be possible by using partitioning information to determine which direction (left or right) to assume for the off-chip routing of the multipliers. LASSIE currently assumes that all off-chip modules are routed off the left of the chip. See [Lagnese89a, Lagnese89b] for more details. A segment of the partitioned design layout produced by LASSIE is shown in Figure 8-5.

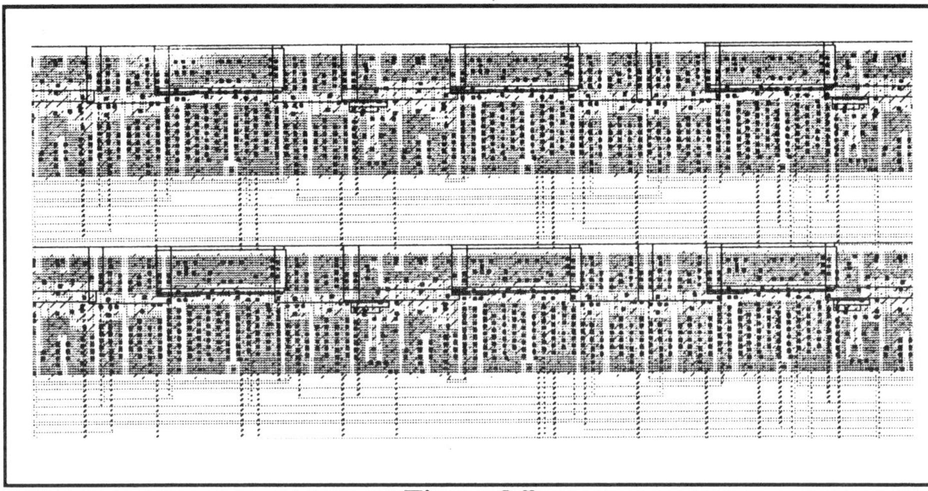

Figure 8-5
Portion of Layout Produced by LASSIE for the Elliptical Filter

8.1.7. Elliptical Filter Summary

Since the Elliptical Filter is a small design, a number of design alternatives could be presented here for comparison, and a number of conclusions can be drawn.

Comparing the vertically-split two-process design (C) to the horizontally-split designs (B1, B2, B3), the vertically-split design requires more microcode than the single-process design, whereas the horizontally-split designs generally require less microcode. However, the vertically-split design requires only slightly more hardware than the single-process design, and no additional functional units, whereas most of the horizontally-split designs require additional hardware.

Another difference between the vertically-split design and the horizontally-split designs lies in the amount of data transferred between the two processes. When the design is split before x16 (Design B2), 6 values are transferred from the first process to the second, and 3 values are returned, assuming that values used exclusively by one process are not transferred. For the vertically-split two-process design, 3 values are sent from the first process to the second process, and 2 values are returned. If each process was implemented on a separate chip, the vertically-split design would require 44% less inter-chip data transfer, possibly reducing the amount of driver circuitry and external pins, although busses might be employed to reduce the discrepancy in the number of pins. The partitioned design requires even fewer inter-partition data transfers, but as explained previously, the Workbench tools currently do not have the ability to make these partitions separate processes, so control information would have to be passed between the partitions in addition to the data transfers.

As stated earlier, these results (reduction in per-process area or microcode) are not guaranteed to occur for all designs, or for all process divisions. However, the idea that the per-process area and the number of values shared between processes can be varied with different process formations follows our intuition. The Workbench, with its behavioral transformations, provides a computer-aided environment in which to explore such alternatives.

The first 3 designs, A, B, and C and the last 2 designs, D, and E vary in the amount of hardware assumed for scheduling, and thus in the amount of resources required for the designs.

By comparing the single process, limited hardware design (D) and the partitioned design (E), the Elliptical Filter demonstrates how partitioning can guide the design process from the behavioral description through the layout of the design. By consistently making use of the structural information provided by the partitioning, results were improved in each stage of the design process. The scheduler was given a preliminary idea of resources available for scheduling so that it could make a schedule with the appropriate level of parallelism. Since parallelism was determined appropriately, the data path allocator was able to find a less complex interconnection scheme according to the partitions. The division of hardware provided by EMUCS allowed for a cleaner bussing scheme with more localized busses, and the overall area for the layout was improved. All this is the effect of following a global goal for the design in each phase of the synthesis process. It is important that the same design plan be followed by each of the design tools so that the structural separation discovered by one tool is not overruled by a later tool.

Overall, the Elliptical Filter shows how the first two tools in the general synthesis path can be used to explore the system level design space. The utility of behavioral transformations in exploring various multi-process options was demonstrated. Also, the advantage of using high-level partitions to guide the synthesis task was shown.

8.2. KALMAN FILTER

The Kalman Filter is part of a real time controller. It is presented here for two reasons. First, as a controller, this machine is neither a DSP machine nor an instruction set architecture, so it demonstrates the breadth of designs that can be addressed by the general synthesis path in the Workbench. Second, this design is small enough to present several design variations which are a result of changing design costs to allow for design space exploration.

The Kalman Filter ISPS description is given in Figure 2-3. The VT for this architecture is substantially different from the Elliptical Filter VT in several ways. First, the Kalman Filter has multiple vtbodies, rather than a single basic block. The multiple vtbodies are a result of the looping constructs in the description. Also in contrast to the Elliptical Filter which has no explicit control specified, the looping and decision constructs in the kalman ISPS produce control operators in the Kalman VT. The Kalman Filter also has array structures defined, which become memories in the register transfer designs produced by the Workbench. There are 73 operators in the Kalman Filter VT. To create a design, the VT was run through APARTY, CSTEP, and EMUCS/Busser. Behavioral transformations were not used because the vtbodies cannot be expanded in-line because they represent loops. In EMUCS, the design space was explored by using various input cost combinations to produce several different designs.

8.2.1. Partitioning

A data partition was desired for the Kalman Filter in order to reduce global interconnect, so the data partitioning configuration described in Chapter 4 was used. The Kalman Filter was partitioned using 2 stages of data clustering and one stage of operator clustering. The result is an address-data partitioning scheme for the architecture, with the array indexing operations performed in the address partition and the actual matrix multiplication performed in the data section.

8.2.2. Scheduling

The partitioned VT was fed into CSTEP to obtain a control schedule. It was specified that no more than 1 MINUS, LEQ, or PLUS operation should be scheduled at a time per partition under the assumption that these operations are similar enough to be implemented by the same hardware in the target technology. MULT operations were assumed to be implemented by separate hardware. Therefore a MULT could be scheduled in parallel with any other data operator.

Partitioning does not have a large effect on the scheduling of this modular design. The partitions indicate that a PLUS operation may execute in the address partition at the same time that another PLUS operation is executing in the data partition, but it is never necessary to schedule simultaneous PLUS operations, so the partitions do not alter the schedule.

8.2.3. Data Path Allocation

As is the case for scheduling, partitioning does not have a large effect on the data path allocation for this design. This is a result of EMUCS' inability to bind objects of different bitwidths to the same module. In the Kalman Filter, the addressing carriers are 4 bits, while the data carriers are 16 bits, so address and data operators will not share hardware. For this case, this restriction to the way that EMUCS can bind exactly reflects the partitions produced by APARTY.

Results for 3 different runs through EMUCS for the one Kalman Filter schedule are listed in Table 8-5 as Kalman #1, Kalman #2, and Kalman #3. The different runs show the effect of varying input costs to reflect different target technologies.

Kalman #1. Column 1 of the data shows the results of using the same assumptions about the target technology as were used in creating the schedule: MINUS, LEQ, and PLUS are similar, and can be implemented by a single hardware unit. Three functional units were chosen for this data path. In this design, the MINUS and LEQ operations are implemented by a single functional unit, while the PLUS operations are implemented by another functional unit. The PLUS operations are not implemented with the others because they are in a different partition, as shown in Figure 8-6, where partitions are indicated by thick shaded lines. The third functional unit is a multiplier, which is in the same partition as the adder, but costs deemed that it was cheaper to implement a separate functional unit than to try to implement the MULT and PLUS operations in the same functional unit.

The address partition is shown at the top of Figure 8-6 and the data partition is shown at the bottom. The sequencer and the small memories are not considered to be part of any partition, while memories ANEW and K each make up partitions of their own. The quality of this partitioning scheme for the Kalman Filter can be seen in the interconnect of the data path. There are no direct connections between the address partition (at the top of the figure) and the data partition (at the bottom of the figure). All connections are via the memories and busses that sit between the two partitions.

Kalman #2. For column 2 of Table 8-5, the costs of putting the MINUS and LEQ operations together were increased, causing an additional functional unit to be allocated. The higher costs indicate that it is expensive to implement those operations together in the target technology. The resulting design reflects the new costs: the MINUS and LEQ operations are implemented by separate functional units. The additional functional unit has the effect of reducing the interconnection complexity slightly: there is one fewer multiplexor input for the Kalman #2 design than is required for Kalman #1.

Kalman #3. Keeping the same functional costs as for Kalman #1 so that 3 functional units are again allocated for the design, the cost of adding registers was reduced to produce the design in column 3 of Table 8-5. In the two previous designs, the cost of adding a new register was high, causing the minimum of 7 registers required by the schedule to be used. By assuming a different target technology with less expensive registers, a 12 register design was produced, as shown in column 3. By allowing more registers, the interconnect complexity could be improved relative to Kalman #1, as indicated by the number of multiplexor inputs required for the Kalman #3 design.

Clearly the input design costs affect the design produced by EMUCS. To choose appropriate costs, the designer must determine relative costs of design units such as registers and multiplexors for his particular target technology. Specific information about the target technology would allow for a discussion of the relative merits of the various

Table 8-5. Design Units for Various Kalman Designs

Design Unit	Kalman #1	Kalman #2	Kalman #3
Functional Units	3	4	3
Wiring Units (concat)	1	1	1
Registers	7	7	11
Multiplexors	8	8	8
Multiplexor Inputs	21	20	20
Busses	4	4	4
Bus Inputs	10	10	10
Bus Outputs	9	9	9

Kalman designs. For example, if the increase in area implied by the 4 additional registers used in Kalman #3 is smaller than the increase in area implied by the greater interconnect complexity in Kalman #1 for a given target technology, then Kalman #3 is the better design for that technology.

8.2.4. Layout

While this design is small enough to be run through LASSIE, several of the modules used in the data path are not currently available, so the resulting layout cannot be analyzed properly. However, using a Seattle Silicon's ChipCrafter system for completion, a Kalman Filter design produced by the Workbench has been completed through layout. Gate level simulation of part of the design using extracted circuit parameters demonstrated functional and timing correctness. For more details, see [Murgai89] and [Un89].

8.2.5. Kalman Filter Summary

This machine has shown how input costs can be varied in EMUCS to allow for a design space exploration and to allow particular technologies to be targeted in the data path allocation phase.

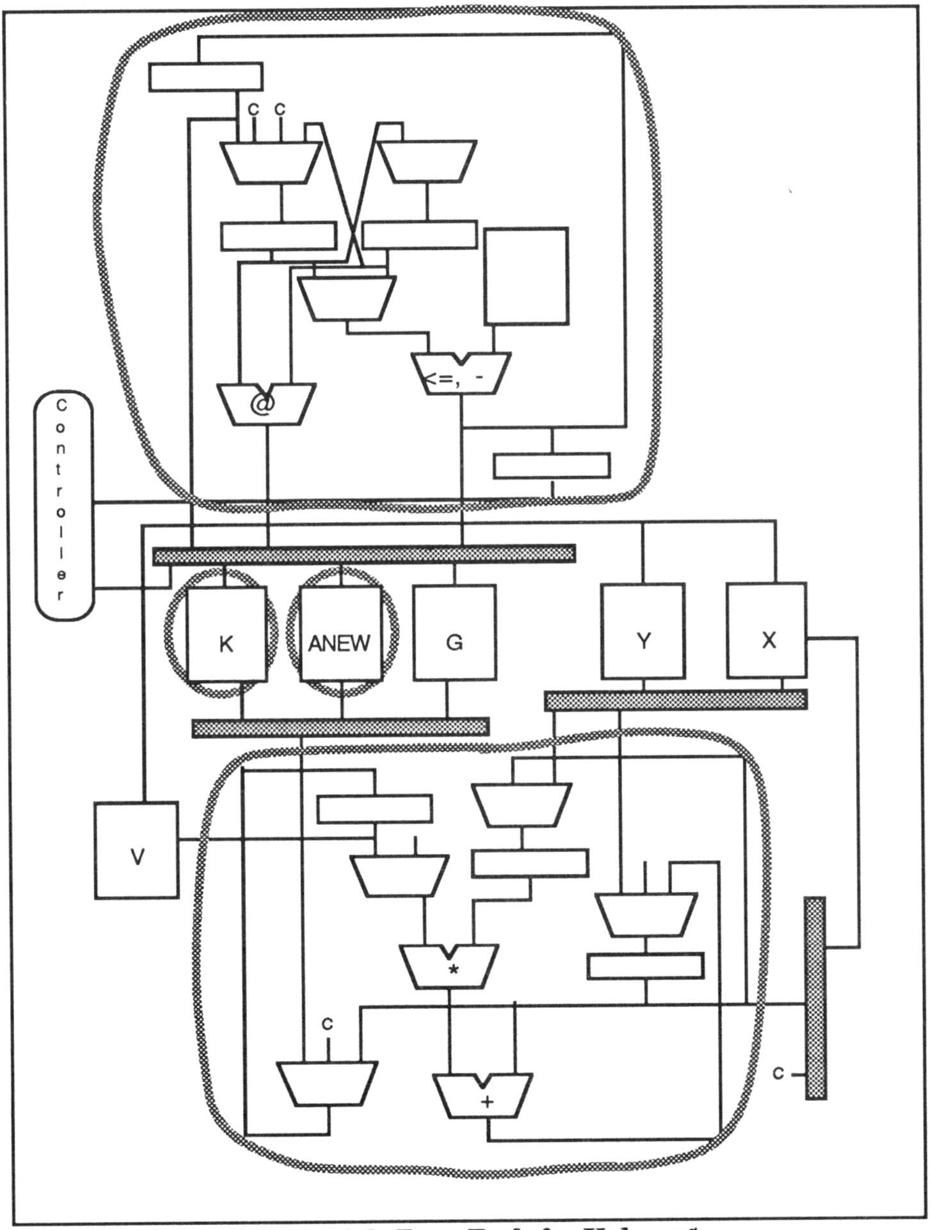

Figure 8-6. Data Path for Kalman1

8.3. BTL310

The BTL310 is a DSP machine designed by Bell Labs. A block diagram of the design is shown in Figure 8-7. The Bell Labs design for the architecture has 6 partitions that can all operate simultaneously: a ROM, a RAM, an address unit (AU), a data unit (DU), an I/O unit, and a control unit (CU).

The VT for this machine was extracted directly from the ISPS, and was run through the partitioning, scheduling, and data path allocation tools.

8.3.1. Partitioning

This description is similar to that of the Kalman Filter in that it has memory operators, control operators, and multiple basic blocks. Therefore the clustering stages are similar for both designs. Two stages of data clustering were used to partition the BTL310. The operator clustering stage used to reduce the overall size of the Kalman partitioning scheme is not necessary for the BTL310 because the partitions produced by the data clustering alone produce a small enough design.

The partitioning produces 4 partitions: ROM, RAM, address calculations, and data execution. This partitioning scheme for the BTL310 matches the partitions chosen by the original designers. The only partitions defined by the designers that do not have corresponding partitions among those chosen by APARTY are the I/O unit, whose behavior is not described in the ISPS for the BTL310, and the control unit. The control operators are grouped with the address operators because there is a large degree of interaction between the program counter, instruction register, and the address calculations. EMUCS, however does not place the controller module within any partition, so the data path produced by EMUCS matches the original design even more closely.

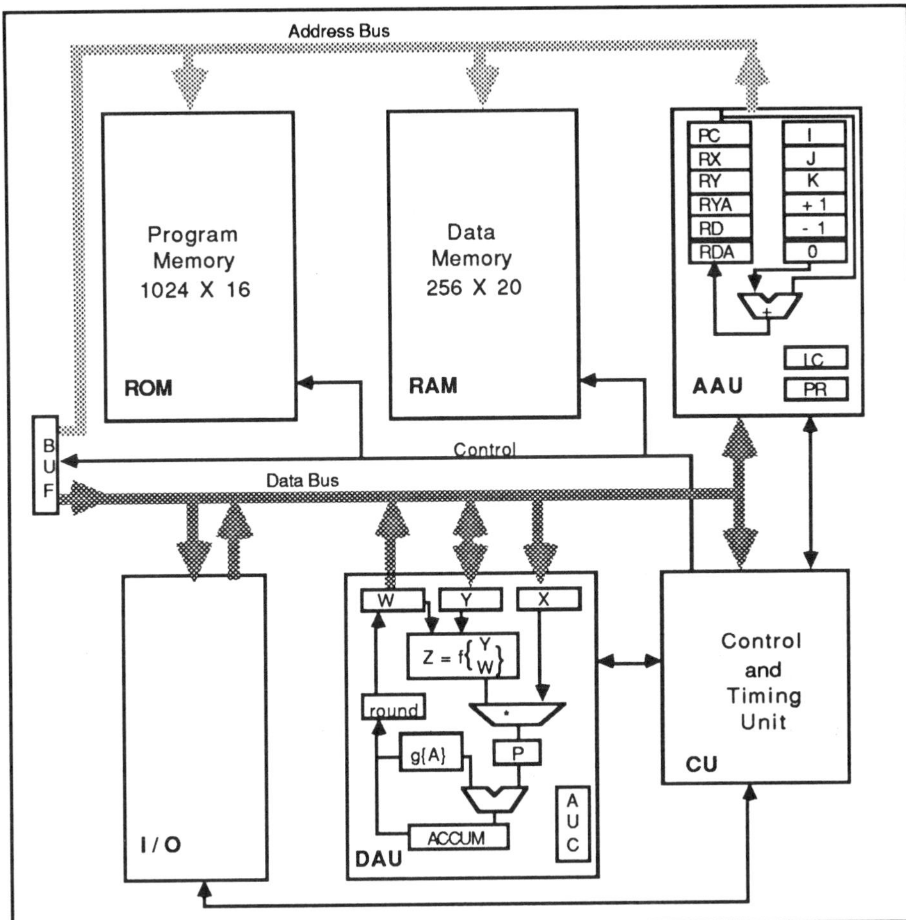

Figure 8-7. Bell's BTL310 Design

8.3.2. Scheduling

Scheduling is straightforward for this design because of the small basic blocks that make up the data flow.

8.3.3. Data Path Allocation

The BTL310 partitions can be run through the data path allocation and bus choosing tools. Module counts for a data path design by EMUCS and Busser using default cost values are shown in Table 8-6.

Table 8-6

Summary of Data Path Allocation Results for BTL310

Design Unit	Quantity
Memories	3
Functional Units	5
Wiring Units	0
Registers	14
Multiplexors	10
Multiplexor Inputs	27
Busses	2
Bus Inputs	9
Bus Outputs	9

8.3.4. Busses

The interconnection scheme between the partitions is shown in Figure 8-8. Two busses provide the wiring for the majority of the interconnections between partitions. These busses correspond to the address and data busses defined in the commercial design. One additional connection is necessary in the design produced by the synthesis tools to address the RAM because the time of its use conflicts with that of both busses. The obvious choice would be to put the RAM address on the address bus, but there is at least one time when the ROM and RAM are addressed simultaneously. The design would have to be re-scheduled to correct for this. By considering the design of Figure 8-8 as a first pass at a final design, information about the inter-partition bussing scheme could be fed back to the scheduler to suggest bus resource limitations. Without the global interconnection phase that is made possible by partitioning, Busser chooses an interconnection scheme that requires 8 inter-partition busses. Since inter-partition busses are assumed to be the longest in the design, the reduction of connections from 8 to 3 as achieved through global considerations appears to be a major improvement in the design. The actual improvement in terms of total wire length for the design can only be determined through layout including appropriate sizing of bus drivers.

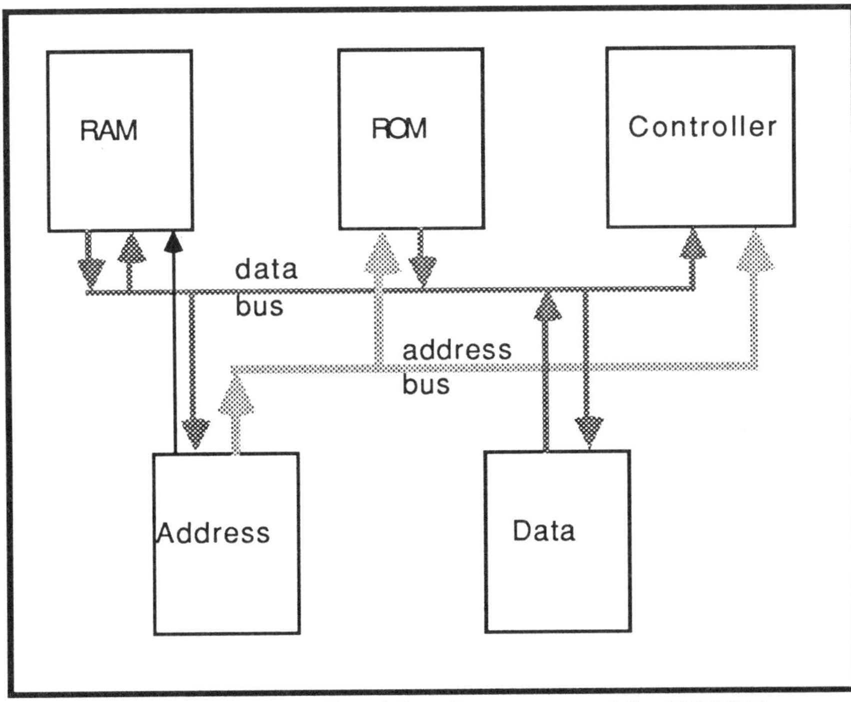

Figure 8-8. Inter-Partition Interconnect for BTL310

8.3.5. BTL310 Summary

The data partitions automatically chosen by the partitioner match the original designers' partitions very closely. By reducing data interconnections, APARTY has extracted the high level structure of the BTL310 design. This information aids the synthesis process in producing a better design with less global interconnect. The overall design matches closely with the commercial design.

8.4. MCS6502

MOS Technology's MCS6502 is a small microprocessor. Like the Elliptical Filter, this architecture was one of the examples from the ACM / IEEE 1988 Workshop on High-Level Synthesis. This section compares design results for both the general synthesis path through the Workbench, and the style specific path embodied in SUGAR with the commercial design.

8.4.1. General Synthesis Results

In the general synthesis approach, behavioral transformations, architectural partitioning, scheduling, and data path allocation are all used for this machine. Four designs are considered here, and are referred to as designs A, B, C, and D. The first, design A, uses no transformations. Design B uses some transformations but is a single-process design. The remaining two designs are behaviorally transformed into multi-process designs. Design C has a 4 stage pipeline, while design D has 2 stages. Only the second of the four designs was completed through the general synthesis path because it can be partitioned, scheduled, and allocated with the current tools. For the other designs, only intermediate results are given.

The initial ISPS description of the 6502 is highly modular, containing many vtbodies, many CALLs, and three levels of instruction decoding. This makes the description easily readable, but difficult to synthesize well, so behavioral transformations are appropriate.

Behavioral Transformations. This section discusses three variations on the original description. The first of these designs (design B) is "flattened," meaning that all vtbodies are expanded inline and SELECTs are combined where possible. The second and third designs build off of this flattened design to create pipelined data flows. Using the transformations described in Chapter 3, the flattened description is transformed to produce two different pipelined alternatives: a 4-stage pipeline, and a 2-stage pipeline with a control schedule similar to the commercial design.

The first step in transforming the design is to eliminate unnecessary microsubroutine calls. Vtbody Inline Expansion can be used to traverse the vtbody call graph recursively, expanding copies of all possible vtbodies inline. After copies of all vtbodies have been expanded inline, the original vtbodies are no longer referenced and can be removed with the Dead Vtbody Elimination transformation, leaving a single large vtbody.

The next step in transforming the design is to reduce the number of levels of instruction decoding. In the original 6502 description, there are three levels of decoding: the instruction group, the instruction opcode, and the addressing mode. Since each decoding operation requires a microcode jump to the appropriate code sequence, and nested decoding operations must be performed sequentially, it may be advantageous to combine nested levels. SELECT Motion and SELECT Combination can be applied, reducing the original description's three levels of decoding to a single level which decodes both the instruction and the addressing modes. The ORA immediate portion of the 6502 VT after flattening to a single level of decode is shown in Figure 8-9.

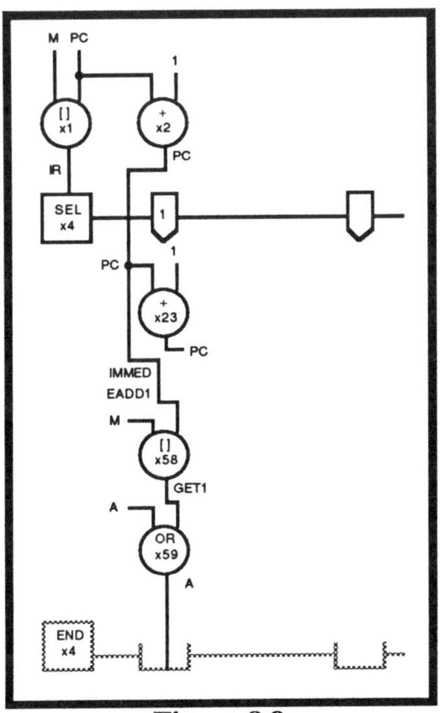

Figure 8-9
ORA Immediate Instruction After Flattening to a Single Level of Decoding

Pipelines can be created for this design using further behavioral transformations; two variations are presented here. The first alternative is illustrated on the left side of Figure 8-10. In this alternative, the design was partitioned into a 4-stage pipeline, with

IFETCH, DECODE, OPFETCH, and EXECUTE stages. The second alternative, illustrated on the right side of the figure, is partitioned more like the commercial MCS6502, and divides the design into a 2-stage pipeline, with an IFETCH stage and a DECODE / OPFETCH / EXECUTE stage. Both alternatives are derived from the same unpipelined ISPS behavioral description, and are formed using the transformations described in Chapter 3. Other alternatives, with different stage divisions, could also be created.

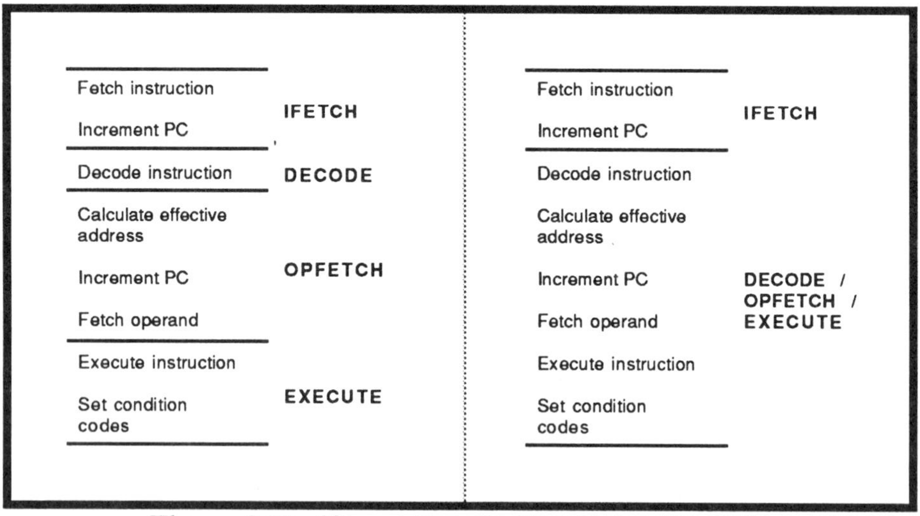

Figure 8-10. Alternatives in Pipelining the 6502

The following outline shows how behavioral transformations can be used to create design C, a four pipestage description, from the unpipelined description. The four stages for this design are: IFETCH, DECODE, OPFETCH, and EXECUTE.

1-2. Proceed as for flattened design, except SELECTs are combined to form 2 levels of nested decoding.

3. SELECT Duplication, SELECT-to-SELECT Transfer, and SELECT Output Closure are applied to form SELECT "stages" and move operators between the SELECTs.

4. The SELECT "stages" are encapsulated into vtbodies, the vtbodies are converted into processes, and the processes are converted into pipestages.

   Steps 3 and 4 pipeline the data flow to increase the instruction initiation rate.

For this example, in step 2, SELECT Motion and SELECT Combination are used to combine the instruction group and opcode decoding into one level, leaving the addressing mode decoding nested to form 2 levels of decoding. Figure 8-11 shows the ORA immediate portion of the 6502 VT after this flattening.

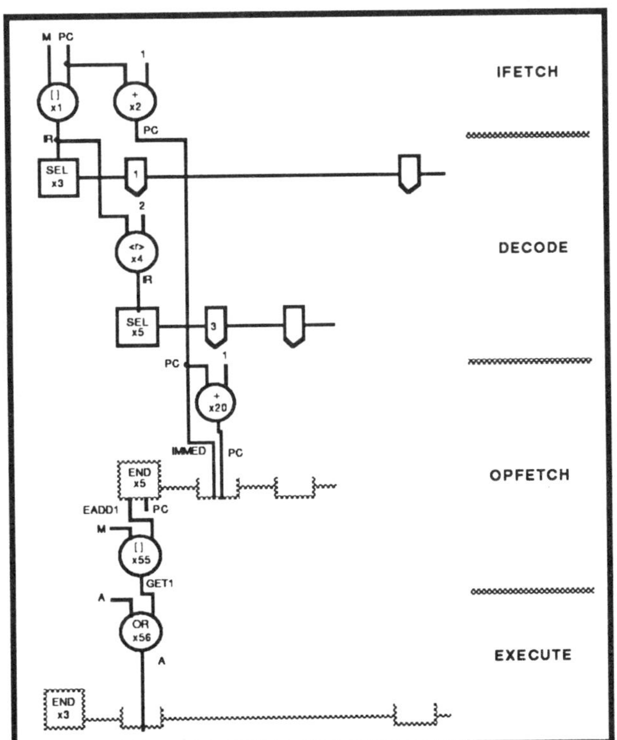

Figure 8-11
ORA Immediate Instruction After Flattening to Two Levels of Nested Decoding

After flattening, the next step in the transformation process is to duplicate the instruction and address mode decoding SELECTs to produce SELECT "stages"; these stages will later be encapsulated into vtbodies and formed into processes and pipestages. In this example, the stages to be formed are the IFETCH, DECODE, OPFETCH, and EXECUTE stages (see Figure 8-11), so the instruction decoding SELECT (x3) must be duplicated twice (once to produce a copy for the DECODE stage, and once to produce a copy for the OPFETCH stage). The original SELECT will become the EXECUTE stage, and the IFETCH stage has no need for a SELECT.

To create the DECODE and OPFETCH "stages", the SELECT Duplication transformation is used twice to duplicate the main instruction decoding SELECT (x3). This results in two new SELECTs (x1064 and x1065) placed immediately before the original in the VT — one of these will become the DECODE stage, and one will become the OPFETCH stage. The results of this duplication are shown in Figure 8-12. The addressing mode decoding SELECT (x5) can be duplicated similarly.

After the instruction decoding SELECT has been duplicated to create SELECT "stages", the operators in each of the 6502's instructions must be moved into the appropriate SELECTs (and thus into the appropriate pipestages). For example, the operand fetch operator (x55) is moved up into the OPFETCH stage, but the OR operator (x56) is left in the EXECUTE stage. Similarly, the duplicate of the addressing mode decoding SELECT (x1066) is moved up into the DECODE stage, and the original addressing mode decoding SELECT (x5) is moved up into the OPFETCH stage. All of the 6502's instructions are processed in this manner. Finally, SELECT Output Closure is applied to clean up the VT. Figure 8-13 shows the result.

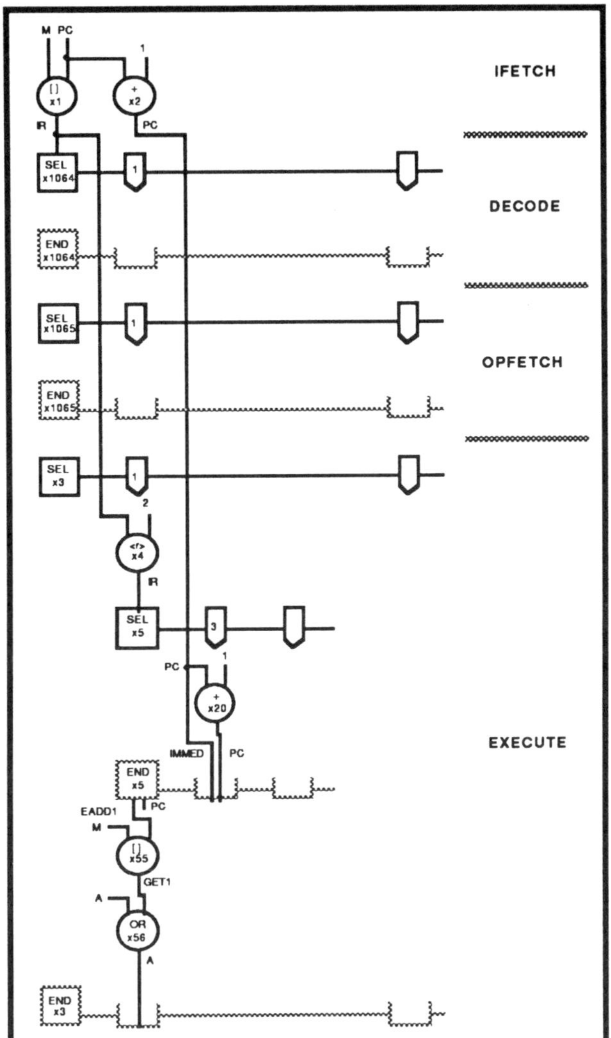

Figure 8-12
ORA Immediate Instruction After SELECT Duplication

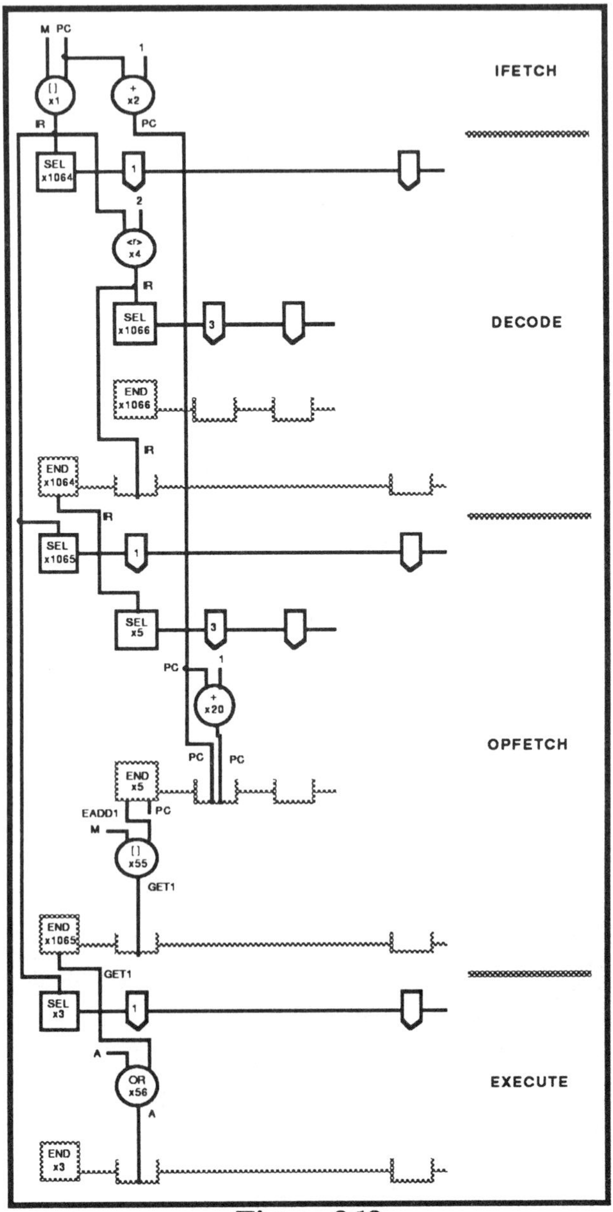

Figure 8-13
ORA Immediate Instruction After SELECT-to-SELECT Transfer for 4 Pipestage Description

The next step in forming pipestages is to place the SELECT "stages" into separate vtbodies, then transform each vtbody into a process and then a pipestage. First, Vtbody Formation is used to encapsulate each SELECT into a separate vtbody. As shown in Figure 8-14, this results in four vtbodies: an IFETCH vtbody, which increments the PC; a DECODE vtbody, which decodes the instruction and addressing mode; an OPFETCH vtbody, which calculates the effective address and fetches the operand; and an EXECUTE vtbody, which executes the instruction.

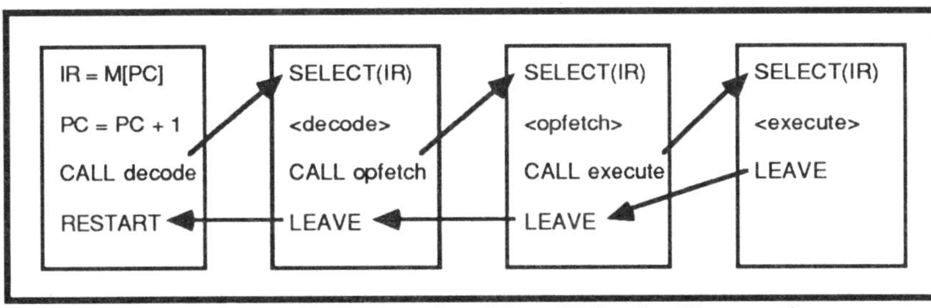

Figure 8-14. Call Graphs After Vtbody Formation

Once the vtbodies have been formed, they can be converted into processes and then into pipestages. Process Creation marks the vtbodies as processes, replaces the CALL and LEAVE operators with SEND and RECV message-passing operators, and RESTARTs the processes so that they will operate continuously. Pipestage Creation removes the intermediate SEND and RECV operators so that the final stage can forward any changed values directly to the initial stage, and adds the assumption that the stages will execute in lockstep and that all data transfers between the pipestages will occur at the same time. The result is shown in Figure 8-15.

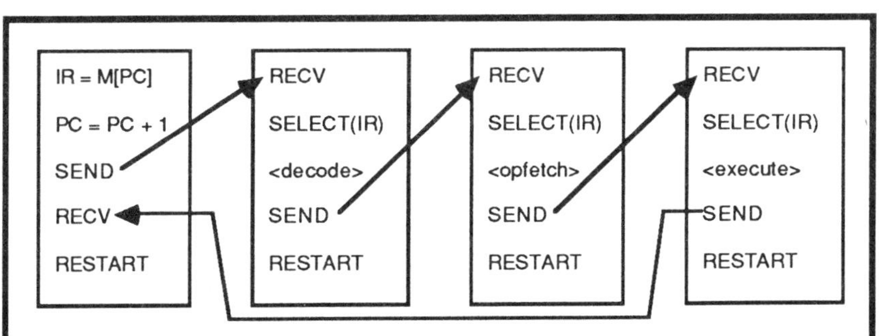

Figure 8-15. Call Graphs After Pipestage Creation

After the pipestages have been formed, SELECT Motion can be applied to the design once again, this time to move the SENDs in the three final stages inside the SELECTs. Since these were often the only operators following the SELECTs, a better control step packing can be achieved after the motion, and the instruction execution time is improved. These transformations complete the development of design C's 4 pipestage behavior.

Beginning with the flattened description (design B), a 2-stage 6502 can also be produced (design D), with an IFETCH stage and a DECODE&EXECUTE stage. As for design B, Vtbody Inline Expansion and the SELECT transformations can be applied to flatten the design into a single level decode. With only these two stages, SELECT Duplication and SELECT-to-SELECT Transfer that were used for the 4 stage design C are unnecessary, so Vtbody Formation is used to encapsulate the instruction decoding and execution portion of the VT into a new vtbody, and Process Creation is used to convert it into a process. Since there are no intermediate stages, Pipestage Creation is also unnecessary. Design D matches the commercial design more closely than the other general path designs, as will be shown in terms of scheduling for the various designs.

Other designs, with a different set of stages, or with the transformations applied differently, can also be created from this same starting description. Thus these transformations allow a designer to explore different pipeline structures, evaluating each in terms of the control schedule.

Partitioning. Pipelined designs cannot be completely synthesized using the current tools, as discussed in Section 3.3, so partitioning was run on the flattened description (design B), which ultimately was completely synthesized through the general path. Data partitioning was chosen as the appropriate scheme for this design to reduce global interconnect. Therefore data and operator clustering were used. Two stages of data clustering exhausted the productivity of this type of clustering: a third stage of data clustering had no effect. Since a smaller overall design size was desired, operator clustering was applied to clean up the operator groups. Four partitions were produced: memory; data; instruction addressing (push, pop, index, PC, and IR operations); and data addressing operations.

Scheduling. Although the Workbench has no data path allocator capable of synthesizing pipelined designs, all 4 VT descriptions can be scheduled. CSTEP is used to assign control steps to the design, and individual instructions are "simulated" to determine their execution times. Additionally, the pipelined designs' control steps are processed partially by hand to take into account pipelined SELECTs, which are not considered by CSTEP. For all designs the functional cost table reflects the assumption that comparison functions are implemented separately from ALU functions.

The first design is the original un-transformed, highly modular description. Although the description is very readable, the control structure is inefficient. For example, with a microcoded controller, each of the CALLs forces a microsubroutine jump, as does each of the SELECT operations. This design requires 16 cycles to execute an ORA (OR accumulator) immediate instruction; this is unacceptably slow, and so will not be considered further for synthesis.

The second design is the flattened description, which was produced by 2 flattening transformation steps. With inline expansion, the CSTEP control schedule shows that the number of cycles required to execute the ORA immediate instruction drops from 16 to 7, a 56% improvement. The SELECT transformations reduce the execution time for the ORA immediate instruction down to 6 cycles.

A simulation trace for the ORA immediate instruction shows the following. The SELECTs in the OPFETCH and EXECUTE stages have been merged by hand into the next cstep under the assumption that the SELECT in the DECODE stage will perform the decoding once, and pass dispatch addresses to the following stages (thus these SELECTs will not require a separate cstep to execute). In this listing, blank lines separate control steps:

```
v46.x1        IR = M[PC]
v46.x2        PC = PC + 1
v46.x3        SELECT (IR)

v46.x4        x4.p1:IR = IR<4:2>
v46.x5        SELECT (x4.p1:IR)

v46.x20       PC = PC + 1
              EADD1 = PC
              IMMED = PC

v46.x55       GET1 = M[EADD1]

v46.x56       A = A OR GET1
v46.x57       x57.p1:SETNZ = A<7:7>
v46.x58       P<7:7> = x57.p1:SETNZ
v46.x59       x59.p1 = A EQL 0
v46.x60       P<1:1> = x59.p1
              SETNZ = A

v46.x1063     RESTART @v46:RUN
```

For the next design, the 4 pipestages reduce the estimated time required to execute an ORA immediate instruction from 16 cycles for an untransformed design to 5 cycles for the pipelined design, and reduces the estimated instruction initiation rate from one new instruction every 16 cycles to once every 2 cycles.

A simulation trace for the ORA immediate instruction shows the following. Since CSTEP does not understand pipelining, the control steps for pipelined SELECTs have been combined with the next control step in the listing, under the assumption that the first SELECT will perform the decoding once and pass dispatch addresses to the following SELECTs. In this listing, blank lines separate control steps, and horizontal bars separate pipestages:

```
v46.x1      IR = M[PC]
v46.x2      PC = PC + 1
v46.x3      SEND {CALLv50} (40)
v46.x4      RECV {LEAVEv50} (38)
v46.x5      RESTART @v46:RUN


v50.x1      RECV {CALLv50} (40)
v50.x2      SELECT (IR)

v50.x3      x3.p1:IR = IR<4:2>
v50.x4      SELECT (x3.p1:IR)
v50.x5      SEND {CALLv51} (43)
v50.x6      RESTART @v50:DECODE


v51.x1      RECV {CALLv51} (43)
v51.x2      SELECT (IR)
v51.x3      SELECT (IR)
v51.x24     PC = PC + 1
v51.x25     GET1 = M[PC]
v51.x26     SEND {CALLv52} (63)
v51.x27     RESTART @v51:OPFETCH


v52.x1      RECV {CALLv52} (63)
v52.x2      SELECT (IR)
v52.x3      A = A OR GET1
v52.x4      x4.p1:SETNZ = A<7:7>
v52.x5      P<7:7> = x4.p1:SETNZ
v52.x6      x6.p1 = A EQL 0
v52.x7      P<1:1> = x6.p1
v52.x8      SEND {LEAVEv50} (38)
v52.x9      RESTART @v52:EXECUTE
```

The commercial MCS6502 is a 2 pipestage design having a prefetch stage and a DECODE & EXECUTE stage. This implementation requires 3 cycles to execute the ORA immediate instruction. In the first cycle, which generally occurs as the previous instruction finishes its execution, the instruction is fetched. The second stage performs the second and third cycles of the instruction. In the second cycle, the instruction is decoded, and at the same time the data at M[PC+1] is fetched (in this case the operand). In the third cycle, the instruction is executed, and a new instruction is fetched. Since this new instruction is being prefetched while the current instruction is being executed, the latency is 2 cycles.

Thus, for ORA immediate instructions, the commercial design which has a pipelined instruction prefetch:

- requires 3 cycles to execute each instruction
- can execute a new instruction every 2 cycles

In contrast, the 4 pipestage Workbench design:

- requires 5 cycles to execute each instruction
- can execute a new instruction every 2 cycles

For the Workbench's 2 pipestage design (design D), the resulting control step schedule for the ORA immediate instruction is shown below. As before, blank lines separate control steps, and the horizontal line separates the two pipestages. As with the commercial design, the ORA immediate instruction requires 3 cycles to execute, and a new instruction can be initiated every 2 cycles:

```
v46.x1       IR = M[PC]
v46.x2       PC = PC + 1
v46.x3       SEND {CALLv47}
v46.x4       RECV {LEAVEv47}
v46.x5       RESTART @v46:RUN


v47.x1       RECV {CALLv47}
v47.x2       PC = PC + 1
v47.x3       GET1 = M[PC]
v47.x4       SELECT (IR)

v47.x30      A = A OR GET1
v47.x31      x31.p1:SETNZ = A<7:7>
v47.x32      P<7:7> = x31.p1:SETNZ
v47.x33      x33.p1 = A EQL 0
v47.x34      P<1:1> = x33.p1
v47.x35      SEND {LEAVEv47}
v47.x36      RESTART @v47:EXECUTE
```

Table 8-7 shows a set of instruction execution times for several ORA addressing modes, comparing the execution times from the 2-stage 6502 CSTEP control schedule to those of the commercial MCS6502. Both designs are pipelined with an instruction prefetch; the extra cycle for the prefetch is not included in the table. In the second column, the number shown in parenthesis includes an extra control step so that no memory references occur in the final control step of the EXECUTE stage; this avoids memory conflicts between the IFETCH and EXECUTE stages which is necessary if a dual port memory is not available. CSTEP assumes single access memories (as is used in the commercial design), but does not recognize an access conflict between the separate processes due to its lack of a global view of the scheduling among processes. As can be seen from the table, the Workbench design is comparable to the commercial design.

Table 8-7

General Synthesis Path 6502 Instruction Execution Times (Cycles)

Instruction and Addressing Mode	2-Stage 6502 Example	Commercial MCS6502
ORA, Immediate	2	2
ORA, Zero Page	2 (3)	3
ORA, Zero Page, X	3	4
ORA, Absolute	3 (4)	4
ORA, Absolute, X	3 (4)	4
ORA, Absolute, Y	3 (4)	4
ORA, (Indirect, X)	4 (5)	6
ORA, (Indirect), Y	4 (5)	5

Data Path Allocation. Since the allocation tools currently cannot build pipelined designs, only the flattened design will be discussed from here on for the general synthesis path. All of the actual functional units for this design are listed in Table 8-8. For each partition, the "Functional Unit" column assigns a name to the type of functional unit. The next column lists the operations performed by each functional unit, and the final column lists the bitwidth of each functional unit.

The data partition requires an ALU, a shifter, and a comparator. The ALU and shifter are separate functional units because they are of different bitwidths. The comparator is separated from the ALU due to the functional costs assumed for this design. The functional costs for these operations were specified to be high so that comparison and ALU functions could execute in the same cstep, as is done in the commercial design. A difference between the general path design and the commercial design is that the commercial design implements the ALU

Table 8-8. Functional Units for General Synthesis Design

Partition	Functional Unit	Functions	Bitwidth
Memory	*None*	*array read, array write*	*8*
Data	ALU	minus, plus, xor, and, or	8
	Shifter	srr, sr0, slr, sl0	9
	Comparator	eql, geq	8
Instruction Addressing	Adder	plus	16
	Adder	plus	8
	Comparator	neq	8
Data Addressing	Adder+	plus, minus	8

comparisons in the same functional unit, but can do both an ALU operation and a comparison in a single clock cycle through multi-phase clocking.

Each of the addressing partitions require adders. The instruction addressing partition requires both an 8-bit adder for indexed addressing and a 16-bit adder for PC increments and branching. The instruction addressing partition also requires a NEQ comparator for branching. The data addressing partition has an 8-bit adder/subtractor (called "Adder+" in Table 8-8) used strictly to calculate data addresses.

Busses. The data path produced by the general synthesis path in the Workbench has 2 inter-partition connections, which is a 71% reduction from the 7 inter-partition connections required without the global bussing phase. The 2-connection scheme is shown in Figure 8-16.

Summary of MCS6502 through General Path. Several design were developed and compared using various tools and design options to explore the design space. The untransformed description, design A, can be contrasted directly with design B, to which a basic set of transformations was applied. The better schedule produced for design B shows the importance of running behavioral transformations so that the VT that is passed on to the other tools is more easily manipulated to produce good designs. Designs C an D show how the behavioral transformations can be used to create pipelined designs. The 2 stage design (design D) is particularly interesting because its schedule matches closely with that of the commercial design.

The following section discusses the results of using the style-specific path through the Workbench to synthesize the same architecture. After this discussion, the SUGAR designs are compared with the completed general path design (design B).

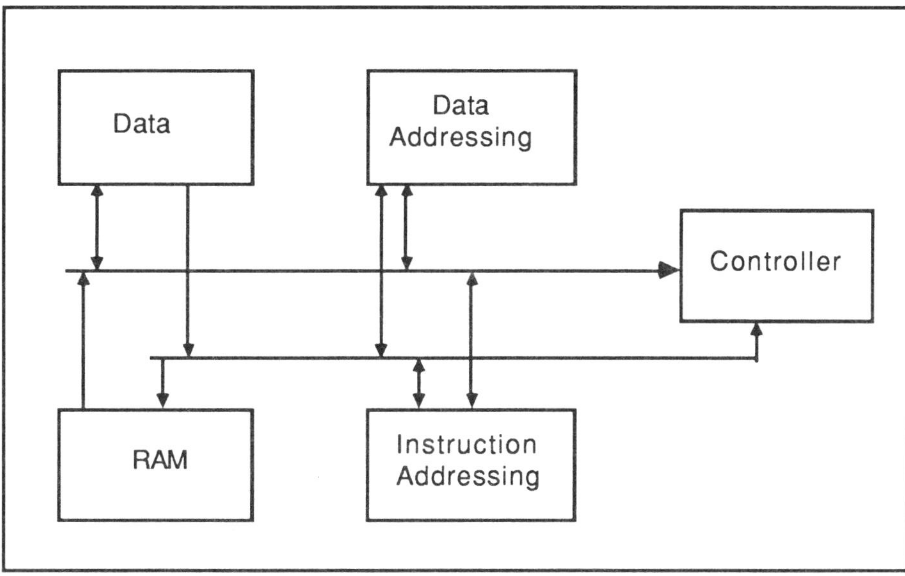

Figure 8-16
Inter-Partition Connection Scheme Chosen by the General Synthesis Tools

8.4.2. SUGAR Results

Three experiments were done with the MCS6502 microprocessor using SUGAR's style specific approach to microprocessor design. These experiments, identified as S1, S2 and S3, are as follows:

1. *S1:* The busses were allocated top-down using the MC68000 style bus model. The design was constrained to have at most one 8-bit ALU and at most two 8-bit incrementors. The address-high section and the address-low section were each required to have at most one 8-bit incrementor and no ALUs. The data-low section was constrained to have no incrementors and at most one 8-bit ALU.

2. *S2:* The busses were allocated bottom-up. The design was constrained to have at most one 8-bit ALU and at most two 8-bit incrementors.

3. *S3:* The busses were allocated top-down using the MC68000 style bus model. The design was constrained to have at most one 8-bit ALU in the data-low section and no ALUs in the address-high and the address-low section.

The data paths of S1, S2 and S3 are shown in Figures 8-17, 8-18, and 8-19, respectively. The instruction unit and the control unit are shown separately in Figure 8-20.

Table 8-9 compares the control costs of the three designs. The number of control points (control for data path) was significantly higher for S2. Most of the additional control points were due to the poor interconnection scheme that resulted in a higher number of multiplexor and bus inputs, each of which requires a control signal. S3 requires 4 fewer control points than S1 since the fewer functional units in S3 resulted in the design having a lower parallelism. S3 also had 49% more control states than S1 and 55% more control states than S2. These figures are again because of the lower parallelism in S3. Although S2 had fewer control states than S1, it had more unique control states than S2. This is because each bus transfer microoperation in S2 had the option of using any one of 8 different busses (2 of the 10 busses are dedicated to memory address and memory data) while each bus transfer microoperation in S1 had the option of using only one of 2 different busses (2 of the 4 busses are again dedicated to memory address and memory data).

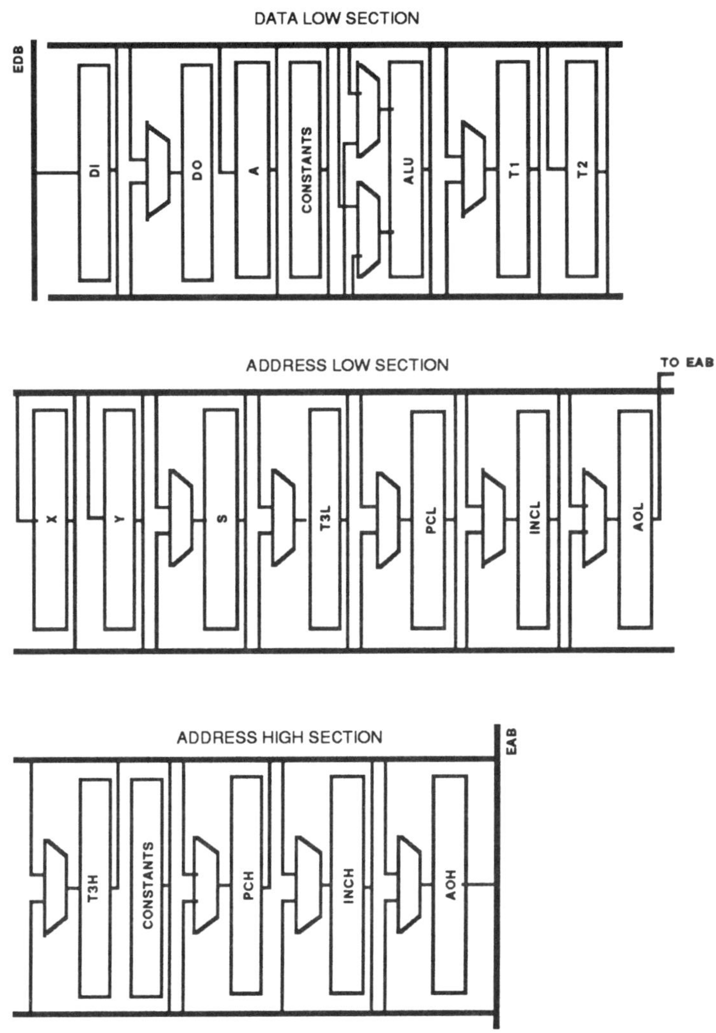

Figure 8-17
Data Paths for MCS6502 S1 Synthesized with 1 ALU and 2 Incrementors Using the MC68000 Style Bus Model

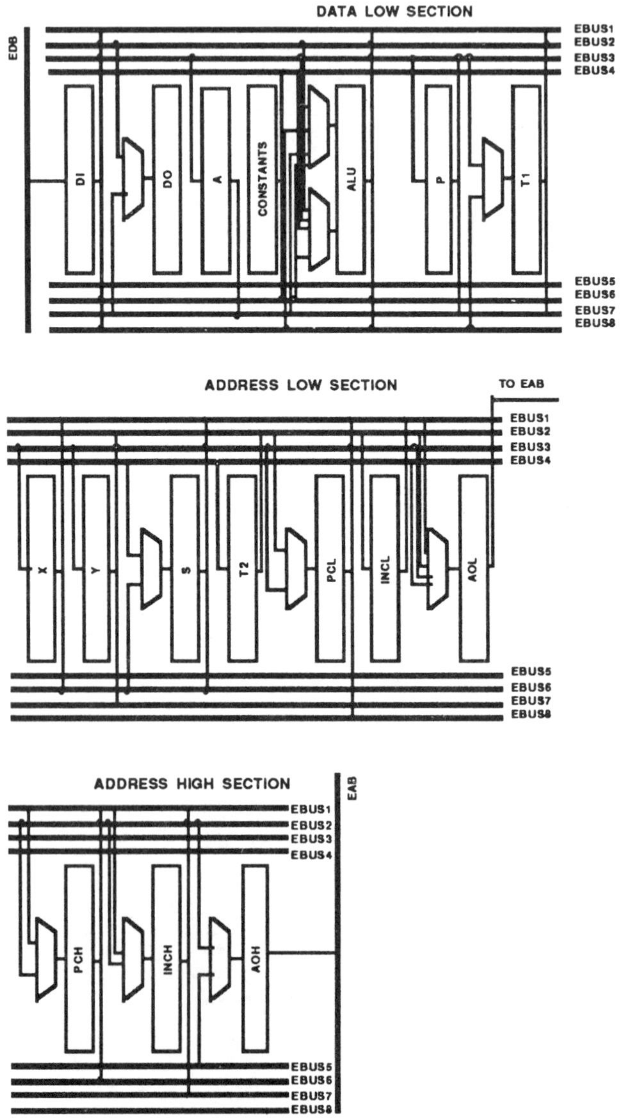

Figure 8-18
Data Paths for MCS6502 S2 Synthesized with 1 ALU and 2 Incrementors Using the Bottom-Up Bus Synthesis Method

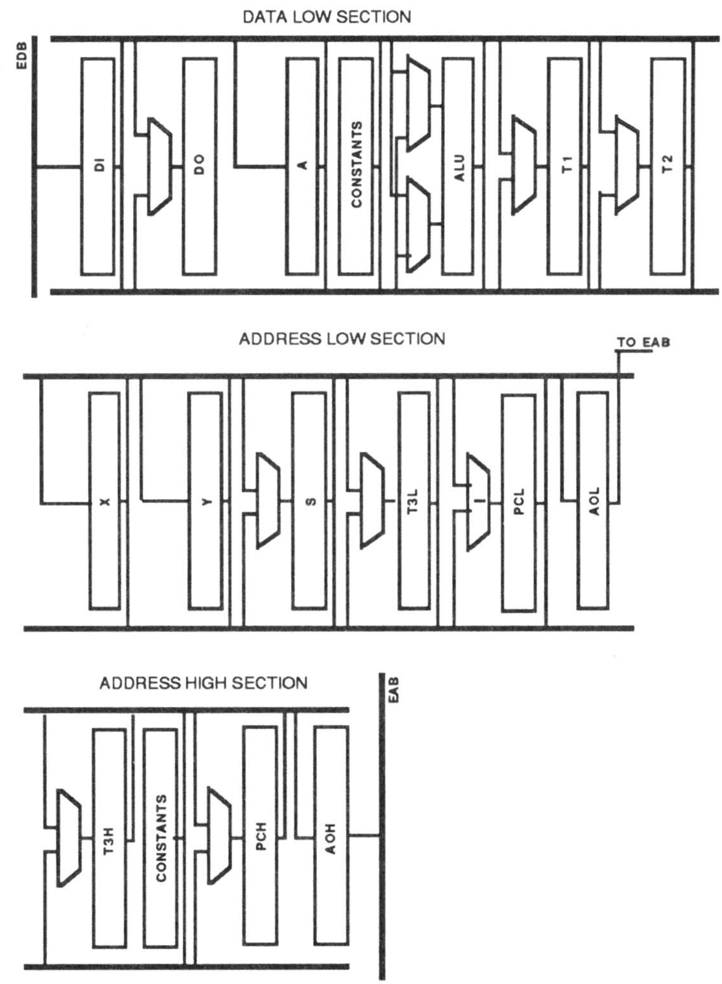

Figure 8-19
Data Paths for MCS6502 S3 Synthesized with 1 ALU Using the MC68000 Style Bus Model

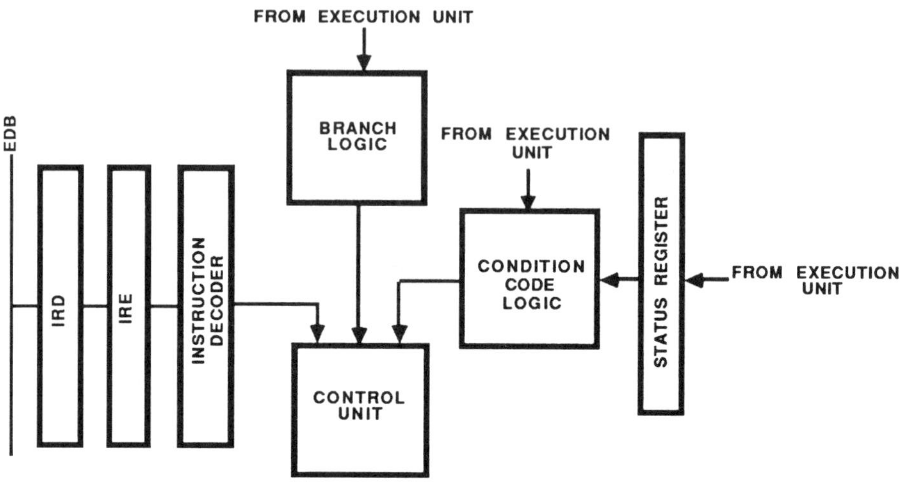

Figure 8-20
Instruction Unit and Control Unit for MCS6502 SUGAR Designs

Table 8-9. Controller Comparison for SUGAR Designs

Measure	S1	S2	S3
Control points	75	91	71
Total states	358	343	533
Unique states	144	150	170

Table 8-10 compares some instruction execution times for SUGAR's designs with the commercial design. All instructions in SUGAR's designs take longer to execute than the corresponding instructions in the commercial design. Instructions with simple addressing modes require an additional 2 cycles while instructions with complex addressing modes require an additional 4-5 cycles. The differences are primarily owing to the inability of SUGAR to prefetch instructions. The branch instructions for SUGAR's designs take 4-5 cycles longer to execute than the branch instructions for the commercial design because instructions are not prefetched and also because, similar to CSTEP, SUGAR uses a local scheduling method which does not allow it to move microoperations past basic block boundaries. A global scheduling method such as trace scheduling [Fisher79] would improve the execution times for the branch instructions.

Table 8-10

Comparison of Some Instruction Execution Times for SUGAR's 6502 Designs

Instruction	S1	S2	S3	Commercial
ADC, IMMEDIATE	4	4	7	2
ADC, ZERO PAGE	5	5	10	3
ADC, ZERO PAGE,X	6	6	10	4
ADC, ABSOLUTE	8	8	11	4
ADC, ABSOLUTE,X	8	8	14	4/5
ADC, ABSOLUTE,Y	8	8	16	4/5
ADC, (INDIRECT,X)	8	8	15	6
ADC, (INDIRECT),Y	10	9	15	5/6
BEQ, RELATIVE	6	5.5	7	2
LDA, IMMEDIATE	4	4	7	2
LDA, ZERO PAGE	5	5	10	3
LDA, ZERO PAGE,X	6	6	6	4
LDA, ABSOLUTE	8	6	11	4
LDA, ABSOLUTE,X	8	8	14	4/5
LDA, ABSOLUTE,Y	8	8	16	4/5
LDA, (INDIRECT,X)	8	8	15	6
LDA, (INDIRECT),Y	10	9	15	5/6
STA, ZERO PAGE	5	5	7	3
STA, ZERO PAGE,X	6	5	9	3
STA, ABSOLUTE	7	6	10	4
STA, ABSOLUTE,X	8	8	13	5/6
STA, ABSOLUTE,Y	8	8	13	5/6
STA, (INDIRECT,X)	8	8	12	6
STA, (INDIRECT),Y	10	9	13	6/7

8.4.3. General Synthesis/SUGAR Comparison

Several general-synthesis-path designs and several SUGAR designs have been presented in this section. Here the results for the three SUGAR designs and the one completed general path design (the flattened design) are compared. Table 8-11 shows a comparison of the area-costs of the Workbench designs with the commercial design. Two other designs synthesized by the DAA and EMERALD systems are also listed. These designs, DAA-2 described in [Thomas83], and EMERALD #1 described in [Tseng84b], were synthesized from a different (although theoretically equivalent) ISPS description. Since many of the basic assumptions about these designs are undocumented, few comparisons can be made between these designs and any of the other designs. However, they show that the Workbench designs are in line with other synthesis results.

Table 8-11. Comparison of Costs of 6502 Designs

Metric	S1	S2	S3	Gen'l	Commercial	DAA	EM'LD
Register Bits	164	148	148	196	—	254†	213
ALU Bits	8	8	8	8	8	17	28
Counter Bits	0	0	0	0	0	—	73
Incrementor Bits	16	16	0	0	16	0	0
Plus Bits	0	0	0	32*	0	17	0
Shifter Bits	0	0	0	9	0	0	9
Comparator Bits	0	0	0	16	0	?	?
Busses	4	10	4	2	—	?	?
Bus Input Bits	320	456	304	98	—	849	758
Bus Output Bits	280	352	256	386	—	241	314
Mux Input Bits	266	322	245	611	—	92	602††
Mux Output Bits	118	118	111	210	—	34	366††
Mux+Bus Inp. Bits	586	778	415	809	-	883	1360
Total States	358	343	533	444(418)	-	941	1360

— Information unavailable or not applicable.
† Includes counter storage bits.
†† Estimated based on number of 2-input multiplexors.
* Includes 1 8-bit adder, 1 16-bit adder, and 1 8-bit adder/subtractor.

All of SUGAR's designs require fewer register bits than does the EMUCS design. The reason that S2 has fewer register than S1 is that some temporary registers in S1 were constrained to certain processing sections of the execution unit and it was not possible to assign them the same physical register even though they did not interfere. Similarly EMUCS is constrained by both bitwidths and partitions. Values must have the same bitwidth and occupy the same partition in order to share a register.

SUGAR's designs have the same number of ALU bits (8) as the commercial design while EMUCS requires more ALU bits. This is partially due to SUGAR's ALU modules which are modelled after actual microprocessor ALUs, using four clock phases to do an ALU operation and necessary comparisons in a single cycle. EMUCS assumes simple ALU modules and the technology input files specified separate comparators to get a shorter schedule. Like the commercial design, S1 and S2 have two 8-bit incrementors whereas DAA and EMERALD use up/down counters instead of incrementors. DAA synthesized some of the addition operations using hardware adders instead of ALUs. Again EMUCS is constrained by bitwidths and partitions. The 9-bit shift operations cannot be included in the 8-bit ALU, and additional adders are needed to supply that functionality to the two address partitions.

EMUCS's bus interconnection costs are the lowest because the busses are run only between partitions. SUGAR's multiplexor interconnection costs are significantly lower than EMUCS's. Clearly there is a tradeoff between the number of bus and multiplexor inputs. The SUGAR designs have the best overall interconnection cost in terms of bus and multiplexor inputs. EMUCS has the fewest busses, which indicates the simplest global interconnection scheme. S1 and S3 each have three 8-bit busses and one 16-bit bus. In contrast, the commercial design has four 8-bit busses. In the commercial design, two of the 8-bit busses double as address busses and execution unit busses while in S1 and S3 the 16-bit bus is used exclusively as a memory address bus. S2 has too many busses and its interconnection cost is high compared to S1 and S3. This suggests that the top-down bus allocation leads to simpler interconnections and lower interconnection costs.

Performance of the general synthesis design falls between that of the SUGAR designs. CSTEP chooses a 444 cstep schedule for the flattened design (although 26 of these csteps are empty and should be eliminated). EMUCS is limited by its inability to synthesize pipelined designs. The quality of the single instruction schedules that were presented previously for the pipelined designs indicates that the potential for a good total schedule exists. However, SUGAR currently has a more complete system for addressing microprocessor designs, both in terms of performance and area.

Overall, SUGAR is much faster than the general path, and produces designs that are closer to the commercial design than the design produced by the general synthesis path. Pinpointing the reason for this is difficult. SUGAR makes use of microprocessor specific information in synthesizing its designs, and thus can target particular bus styles.

When SUGAR allocates busses bottom-up instead of using a predetermined bus structure, it chooses a notably worse interconnection scheme. Thus, style specific information is important to SUGAR's approach to synthesis. However, many of the other techniques used in SUGAR, such as graph coloring and delayed binding of decisions, could also be applied to a more general synthesis task. The contributions of these techniques toward producing better designs is unclear. At the same time, some of the general path techniques might be used to improve SUGAR. Some of the general path tools whose functions are not available in SUGAR bring the general path designs closer to the commercial designs: behavioral transformations allow designs to be pipelined, and partitioning ultimately results in a better interconnection structure.

8.5. MC68000

The MC68000 is a general purpose microprocessor. Its 40 page ISPS is too big for the general synthesis path. Therefore, only SUGAR results are given here for this architecture.

Three experiments were done with the MC68000 microprocessor one of which (*S4*) is described in this section. Descriptions of the other two experiments can be found in [Rajan88].

In *S4*, the busses were allocated top-down using the MC68000 style bus model. The design was constrained to have at most three 16-bit ALUs. The address-low section, the address-high section and the data-low section were each required to have at most one 16-bit ALU. This design and the commercial design can be expected to have similar hardware since they have the same functional units and the same bus model.

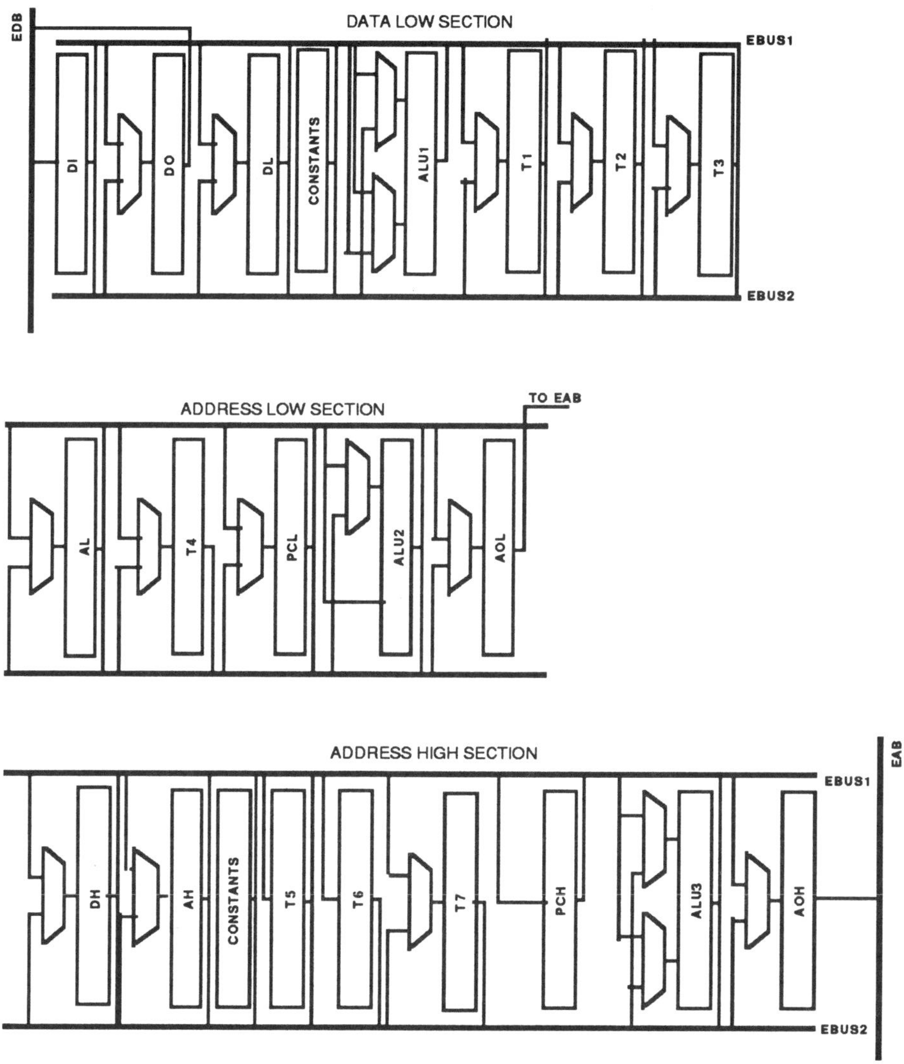

Figure 8-21
Data Paths for MC68000 S4 Synthesized with 3 ALUs Using the MC68000 Style Bus Model

Table 8-12

Comparison of Some Instruction Execution Times for MC68000 Designs

INSTRUCTION	S4	COMMERCIAL
ADD.W Dn, Dn	6	6
ADD.W D, An@(d)	24	16
ADD.L Dn, Dn	8	8
ADD.L Dn, An@(d)	24	24
BRA.W	16	10
BCC.W	16	10/12
MOVE.W Dn, Dn	6	4
MOVE.W Dn, An@(d)	14	12
MOVE.W An@(d), Dn	12	12
MOVE.L Dn, Dn	8	4
MOVE.L Dn, An@(d)	16	16
MOVE.L An@(d), Dn	16	16

Table 8-12 compares some instruction execution times for SUGAR's designs with the commercial design. The instruction execution times for the register to register instructions (*xxx Dn, Dn*) in SUGAR's designs are identical to those of the commercial design or worse by 2 to 4 clock cycles. SUGAR's execution times for instructions with complex addressing modes is higher than those of the commercial design primarily owing to SUGAR's inability to synthesize a design with instruction prefetch. The branch instructions in SUGAR's designs are slower than the commercial design for the same reasons described in Section 8.4.2.

SUGAR's MCS6502 designs used about 8.5M of memory while the MC68000 designs used about 15M of memory. A complete MC68000 could not be synthesized because of UNIX process size limitations. The ISPS of the MC68000 behavior synthesized did not include the privilege instructions, the bit instructions, the subtract instructions, the loop instruction, DBcc and the move multiple instructions.

SUGAR uses a style-specific synthesis method for the microprocessor design style. The method synthesizes even large microprocessors like the MC68000 rapidly in only 80 CPU minutes on a VAX 8800. Although definite conclusions in favor of the style-specific method will require more experimentation, the evidence presented suggests that the microprocessor style-specific method used by SUGAR synthesizes better designs than the general methods used by other synthesis programs.

SUGAR's designs compare favorably with the commercial designs. Most instruction execution times for SUGAR's MCS6502 are slower than the corresponding execution times for the commercial MCS6502 by 2-4 clock cycles and instruction execution times for SUGAR's MC68000 range from being faster than the commercial MC68000's execution times to being slower by 2-10 clock cycles. SUGAR's lack of knowledge of instruction prefetching is the primary reason the instruction execution times for SUGAR's designs are greater than the instruction execution times for the corresponding commercial designs.

8.6. SUMMARY

This chapter has presented synthesis results for a range of architectures with various applications. The designs presented here were chosen to demonstrate the versatility of the Workbench tools in exploring the design space. The Elliptical Filter shows how behavioral transformations can be used to create processes for a design and the physical implications of doings so. The same architecture also demonstrates how the high level structural information extracted by the partitioner can be used to guide synthesis, resulting in a better design. Several designs of the Kalman Filter show how EMUCS can be tuned to reflect various costs of the technology it is targeting. The BTL310 demonstrates several of these techniques for a larger design that has a different application than any of the other designs. The MCS6502 provides common ground for the two synthesis paths through the Workbench, allowing the two approaches to be compared. Finally, the MC68000 shows how SUGAR can be used on a larger design.

9. Correlating the Multilevel Design Representation (CORAL)

Previous chapters have discussed the actual process of synthesizing a structural design from a behavioral description. This chapter offers a method of preserving design information during that process. One of the problems encountered in many automated synthesis systems is that no information is provided describing the relationships between the various intermediate representations of the design that exist in the system. One feeds a design specification into the system and receives an implementation, but the system does not describe the relationship between the input and output design representations.

The CORAL system [Blackburn88a, Blackburn88b] described in this chapter rectifies this shortcoming by gathering and processing information concerning the relationships between the different design representations found throughout the Workbench.

9.1 LINKING DESIGN REPRESENTATIONS

A design system that provides information about the relationships between the different design representations it uses can include *multilevel* design and analysis tools. To support the widest variety of tools, the information should be very specific and have very fine granularity. Ideally, the relationship between two design representations should describe the links between the smallest semantic components common to the two representations. The description of a link should include the type of relationship and the names of the components involved.

If the Workbench is viewed as a black box that accepts Algorithmic Level specifications expressed in ISPS and produces Register-Transfer Level designs, then it should also describe the relationships between the ISPS specification and the Register-Transfer Level implementation. Since operations and values are the smallest semantic elements in an ISPS description, the Workbench should attempt to describe the links between each value and operation in the specification and the corresponding elements in the Register-Transfer Level implementation. This is the goal of the CORAL system.

9.1.1 Linking Scheme

The CORAL system determines the relationships between design representations by gathering information from each of the design tools in the Workbench as they are run. This scheme was chosen to take advantage of the knowledge of the relationships that is present within the tools. The alternative, analysis and comparison of a tool's input and output after the tool has completed, leads to a very complex and expensive solution.

Figure 9-1 shows a schematic diagram of the information flow in one path through the Workbench. The main flow of design information is from left to right through the ISPS to VT translator (Mill), the behavioral transformations, and the register-transfer synthesis phase. Each of these phases of the Workbench records information that describes the links between the operators and values in its input and those in its output. The links from each of these tools are read in by a separate program, called CORAL, where they are joined together to form a chain that reaches from the ISPS input to the Register-Transfer Level output.

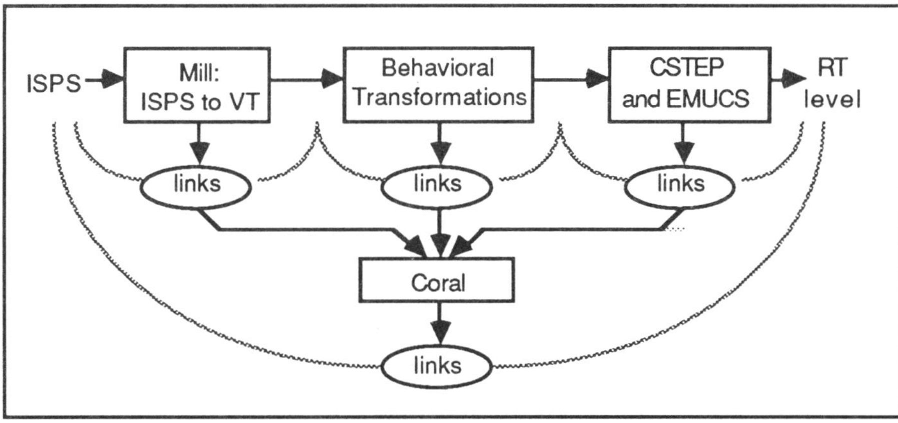

Figure 9-1. The Flow of Linking Information in the Coral System

Each link describes the relationship between two components in adjacent design representations. The designer of such a linking system must specify the relationships that are to be recorded. This must be done for each unique translation or transformation performed by each tool. For instance, for each construct in the ISPS input, the relationships between each component of the construct and the components in the resulting VT construct must be specified. The relationships that must be recorded depend on both the transformations being applied and the requirements of the applications that are to use the linking information. The designer must also specify how the different kinds of relationships combine when joined end-to-end. That is, if component **X** has relationship α to component **Y**, and **Y** has relationship β to component **Z**, then what is the relationship of **X** to **Z**?

At present, the CORAL system only records *equivalence* relationships, recording, for example, that component **X** in the ISPS is equivalent to component **Y** in the VT. Equivalence relations in CORAL are defined to be transitive and commutative. Therefore, if **X** is equivalent to **Y**, and **Y** is equivalent to **Z** in the Register-Transfer Level design, then **X** is equivalent to **Z** and **Z** is equivalent to **X**. There must be an unbroken chain of equivalence relationships between two components in different design representations for them to be considered equivalent.

9.1.2 Implementation

The existing tools in each of the three phases of the Workbench outlined above record the desired information. Future tool specifications should include the recording of this information as part of their original requirements.

The method used to gather the required information depends on the way the particular tool is implemented. Some tools may never modify their output once they have created an initial version of it in memory while others may need to update the structures they create later on. Only the former type can write out the relationships between their input and output design representations incrementally as they run, the rest must have some means of recording the information until all modifications to the design are complete. All of the design tools examined so far are of the latter variety. Of these, some tools destroy their input as they run, either as a direct result of the way they are constructed or as a space saving measure; other tools retain their input in memory until they have completed. Tools of the former type must retain an identifier for each element of the input that must be mentioned in the record of the relationship between the two design representations; tools of the later type may rely on pointers between the two data structures. The Workbench tools that have been integrated into the CORAL system fall in both of these catagories. The mechanisms used to record linking information in each of the three phases described previously are outlined below.

ISPS-VT Translation. ISPS to VT translation is performed by Mill. Mill falls in the category of tools that modify their output representation after initial creation in memory and discard their input representation as they execute in order to save memory space. Consequently, Mill must retain an identifier for every element in the input design description that is to be mentioned in the statement of the ISPS-VT relationship.

In order to describe the relationship of every value and every operator in the ISPS to the corresponding element in the resulting VT, Mill must, in general, retain information about every token in the ISPS file it reads in. This is done by constructing a tag for each token that gives the token's file number (ISPS allows included files), line number within the file, and token number within the line. This serves as a unique identifier for the token.

Each token in the ISPS may represent an operator and/or one or more values. Obvious examples of tokens that represent operators are "+" and "EQL". ISPS operations may also be represented by identifier tokens as, for example, occurs when a CALL operation is represented by the identifier for the CALLed entity. It is also the case when the identifier represents a carrier that is to be read or written. In this case, the identifier may implicitly specify operations that are required to read or write the carrier. As an example, consider the case of a write to an entity which is mapped onto a subfield of a second entity. The identifier implicitly specifies that a bit write operation be performed into a subfield of the target entity.

A single token in the ISPS may also represent one or more values. An identifier on the left hand side of an assignment is said to represent the value assigned. Similarly, an identifier in an expression is said to represent the value returned by the carrier read operation so specified. Furthermore, the token specifying a data operation represents the value returned by the operation as well as the operation itself. In the case of a call operation, all the values returned by the subroutine are represented by the token that gives the called entity's name. The values may be differentiated by specifying the carrier in which they are returned as well as the token representing the call. All the initial values in a block are represented by the first token in the block, typically a BEGIN or "(". The values output from a block or compound statement such as a DECODE are represented by the last token, typically an END or ")".

As a VT is constructed in memory from the ISPS, the identifying tags are copied from the tokens they identify to the resulting constructs in the VT. The tag identifying a "+" token, for instance, is copied to the "+" operator created in the VT to represent it. The tag is also copied to the VT value that represents the output of the "+" operation in order to indicate that this value corresponds to the value returned by the "+" operator in the ISPS. In the case of a call operation, the resulting CALL operator is given the tag from the identifier representing the ISPS call operator. Each of the VT values output from the CALL operator are also given a copy of the identifier's tag supplemented with the identity of the carrier whose value the VT value represents. Tags for ISPS values that do not explicitly exist in the VT, *e.g.* initial values for carries that are not referenced in the current block, are inserted in a special list in the parent VT construct.

After the ISPS-VT compilation is complete, Mill calls a routine to write out the linking information. The routine traverses the VT data structures and, for each operator and value it finds, it writes out the VT operator or value ID along with the ISPS IDs stored in the attached tags. Tags for ISPS values that do not appear in the VT are written out without a paired VT value identifier. In this way, Mill produces a record of the relationships between the values and operations in the two design representations. The record covers all the operations and values in both representations regardless of whether they are linked to a corresponding item in the other.

VT Transformations. The Workbench includes a package of transformations that may be applied to the VT in order to improve the quality of the resulting design (see Chapter 3). Running a set of transformations produces a new VT that will be used by the next tool in the Workbench. Thus, another link in the chain between the original ISPS and the final Register-Transfer Level design representations must be recorded by the transformation package. In this case, the package records a link between two different VT design representations.

By and large, the VT transformation package does not create new VT data structures in memory as it executes. Instead, it modifies the existing structures, changing pointers between nodes in the graph and creating and destroying a few nodes as necessary. Slight changes to the topology of the VT graph, though, often result in wholesale changes to the identities of the nodes. The package needs to record the relationships between the nodes in the new graph and those in the old graph in terms of their new and old identities. It also needs to note the disappearance of those nodes that have been destroyed and record whatever it knows about the relationship of newly created nodes to nodes in the original graph.

The transformation package uses three steps to record this information. First, before transformations are begun, a routine is called to traverse the VT data structure, construct a tag containing the identity of every operation and value in the VT, and attach each tag to the named operation or value. The tags are manipulated as necessary by the transformation routines as they alter the VT data structures. For example, if an operator is duplicated in the VT, the original operator's ID tag is copied to the new operator to indicate that it was derived from the original operator named by the tag. If a VT operator is destroyed, its

ID tag is moved to a special list of tags attached to the operator's parent vtbody.

When the transformation process is complete some operators and values will have been created and some will have been destroyed. Even slight changes to the topology of the VT will have caused many of the operators and values that were not otherwise affected to acquire new names. A routine is called to traverse the new VT data structures and record the identity of every operator and value it finds paired with the identity found on any tags attached to that operator or value. The IDs on tags attached to a vtbody's list of destroyed operators are recorded without being paired with a new VT identifier. In this way, the transformation package is able to record information about the relationships between all the operators and values in the transformed VT and those in the original. This information constitutes the link spanning the VT transformation package in Figure 9-1.

Control Step Scheduling and Data Path Allocation. Control step scheduling (CSTEP) and data path allocation (EMUCS) are the last of the major phases of the Workbench shown in Figure 9-1. The control step sequencing and data path allocation tools synthesize modules to perform the operations and transfer and store the values it finds in the VT. They assign all the operators in the VT to control steps and bind individual values and operators in the VT to specific modules in the synthesized structure. Only the EMUCS data path allocation tool has been integrated into the CORAL system.

To provide the link that spans the synthesis stage in Figure 9-1, EMUCS must record the bindings it makes. It must record the functional unit that each operator is bound to, the register that each value is stored in, the interconnect module (multiplexor or bus) that each value passes through, and the control steps during which these events occur. EMUCS falls into the catagory of tools that create new data structures without substantially altering the original ones; hence, it can rely on pointers to store linking information during its execution.

CSTEP writes the control step each VT operator is scheduled for in the data structure representing the operator. When EMUCS binds an operator to a specific functional unit, it adds a reference to that module in the operator's data structure. The routine used to record VT to Register-Transfer Level linking information needs only to traverse the lists of operators in the VT data structures after synthesis is complete and record the module and control step for each.

VT values are bound to registers and interconnect modules. Since values are not scheduled as are operators, and module references are not written in their data structures by EMUCS, a different mechanism must be used to record the VT to Register-Transfer Level linking information for values. Whenever EMUCS binds a value to a register or interconnect module, it writes a reference to the VT value and the relevant control step in a list attached to the module. The routine used to record VT to Register-Transfer Level linking information examines all the registers and interconnect modules and records the identities of the values bound to the module and the control step at which the value enters it.

The information recorded about operator and value bindings to the structure provides the link spanning the EMUCS and CSTEP phase in Figure 9-1.

Chaining. The final step in the CORAL system is to gather the information that makes up the links spanning each phase of the Workbench and combine it into a coherent statement of the relationships that exist between the initial ISPS representation and the resulting Register-Transfer Level design description.

Since the output of one tool is the input to the next, the components on the output side of the relationship described by one tool exactly match the components on the input side of the relationship described by the next. Hence, the relationships can be joined smoothly to form a chain reaching from the initial ISPS to the final Register-Transfer Level design description.

The CORAL program was constructed to read in the link files written by the tools in the Workbench and join them into such a chain. Link files are read in the order in which they were created. As each file is read, CORAL adds nodes to its internal data structure to represent components in the output design representation described by the link file. These nodes are linked to nodes in the data structure that represent the previous design representation according to the relationships described in the link file.

A further step is taken when VT to VT link files are read. Since transformations may be applied any number of times, there may be a large number of VT to VT link files in the chain between the ISPS and Register-Transfer Level design representations. CORAL avoids storing multiple versions of VT design descriptions in memory by transferring all ISPS links to nodes in the original VT to the corresponding nodes in the transformed VT and discarding its representation of the original VT description.

A simplified version of CORAL's data structure is depicted in Figure 9-2. The nodes labelled 'o', 'v', 'b', and 't', represent operations, values, VT bodies, and ISPS tokens, respectively. Rounded boxes represent groupings. The dark lines and boxes depict links between design representations (a component is linked to a group of components in the next representation, hence the arrows depicting a single link to a group of objects). The grey lines and boxes depict the structure of the design representations themselves. Information about the structure of the design representations is derived from the structure of the link files and is used by some of the application programs described below.

An ISPS representation, as shown here, may be made up of a number of files, each file is a list of lines, each line is a list of tokens, and each token may represent an operation and a set of values. A VT representation is made up of a set of control blocks, each of which consists of a set of operations, and each operation has one or more output values. The structural portion of a Register-Transfer Level representation is made up of a set of modules, each of which has an associated set of values and operations that the module handles and performs.

The diagram in Figure 9-2 shows the links that might be attached to a "+" token in a statement like "A = A + B" where A and B are carriers with different bit widths. In this case, the "+" token represents an addition operation and the value that results from the addition. The operation is linked to two operations in the VT, one which performs the corresponding addition operation in the VT and one, a PAD operator, which extends the value of the smaller carrier to match the width of the larger. The ISPS value represented by the "+" token is linked to the VT value output from the VT "+" operator. The value output from the PAD operator in the VT is not linked to anything in the ISPS. The PAD operator and its output value are linked to the corresponding operator and value associated with a module in the synthesized structure. The

"+" operator and its output value are similarly linked to another module in the structure.

Even in this very simple example there is a one-to-many relationship between an ISPS operator and some Register-Transfer Level operators and there is a VT value that is linked to a Register-Transfer Level value, but does not correspond to any value in the ISPS design representation. These are very simple examples of the complexity of the links that must be maintained as well as the addition of information (operators and values) synthesized in the Workbench.

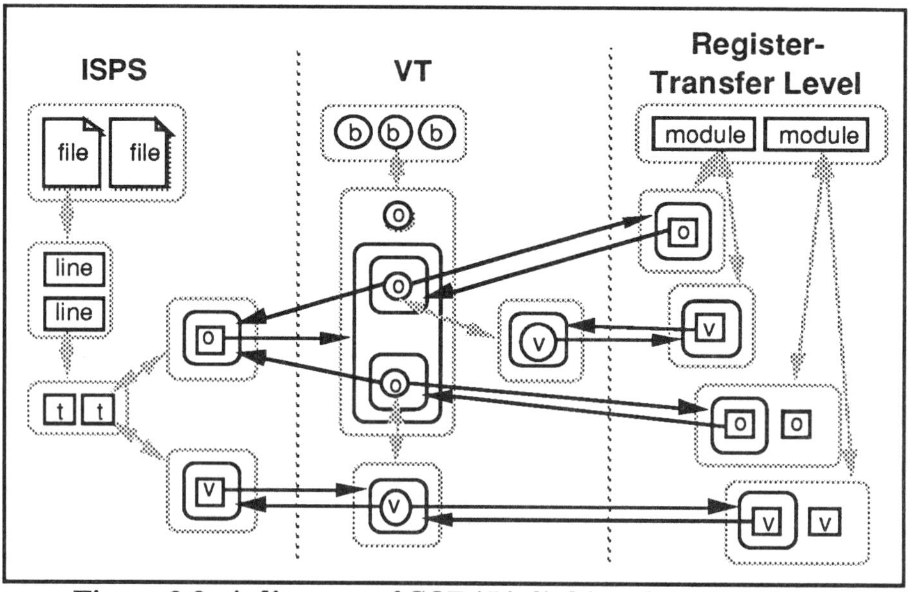

Figure 9-2. A diagram of CORAL's linking data structure.

9.2 APPLICATIONS

There is, of course, no point in producing the above linking information unless it can be used. Several tools have been constructed in the Workbench that demonstrate the utility of having linking information. The SEESAW designer's aid and a package of iterative design tools are described below. A brief discussion of some other possible applications follows.

9.2.1 SEESAW

SEESAW is a graphical tool with which a designer can explore the relationships between the operations and values in an ISPS description and a Register-Transfer Level representation synthesized from it [Koenig87]. SEESAW uses the X Window Manager to create and manipulate windows on a workstation screen. It creates four main windows, one for each of the ISPS, structure, and control description files, and one for SEESAW commands. The ISPS and control files are displayed as text as they appear in the raw files with line numbers added to aid in navigation. The structure description is displayed as a schematic diagram.

The primary use of SEESAW is to explore the synthesized design and learn how it implements the behavior specified in the ISPS file. This is an extremely difficult task without CORAL and SEESAW, as the designer must otherwise hand simulate the register-transfer design and compare its function to the ISPS specification to try to find the corresponding operations and values. With SEESAW, the user need only use the workstation mouse to select a token in the ISPS display and the system will highlight the modules in the schematic that perform the operation(s) and/or handle the value(s) represented by that token. SEESAW also highlights the micro-operations in the control step file that cause the modules to perform those operations. Similarly, the user can select a module in the schematic diagram and SEESAW will highlight all the tokens in the ISPS that represent values handled or operations performed by that module. It also highlights all the operations in the control step file that control the selected module. Similar results are obtained by selecting micro-operations in the control step window.

When SEESAW is run on a color workstation, it allows the designer to choose one of six different colors to use to highlight the next selection. This aids in keeping different relationships separate when they are being viewed at the same time. If multiple relationships target the same element in a window only one color will be shown, but the element will be boxed to indicate the overlap. SEESAW provides a command to display the colors of all the other relationships that target that element.

Figure 9-3 shows a snapshot of a SEESAW display. In this example, the designer has used the mouse to select the functional unit procr_7 in the schematic. SEESAW has highlighted the three "+" tokens in the

ISPS display representing the three PLUS operations performed by procr_7. One of the micro-operations that controls procr_7 has also been highlighted in the control step window. The other two micro-operations that affect procr_7 do not appear in the window in this display.

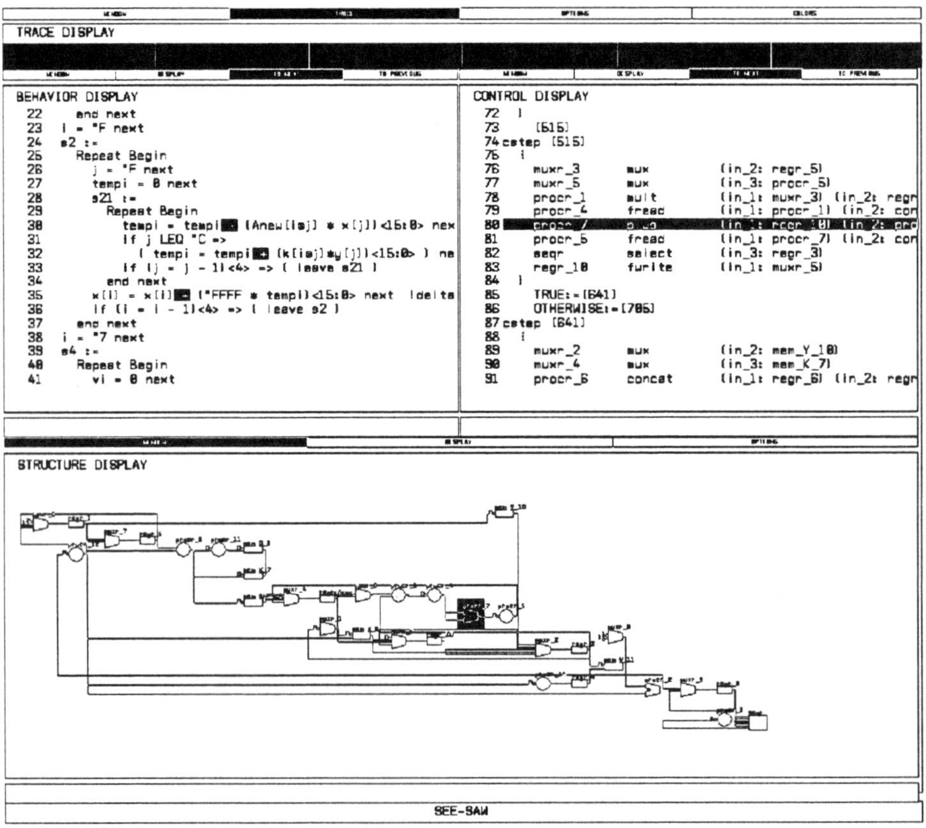

Figure 9-3. A Snapshot of SEESAW Running

SEESAW helps the designer learn how the synthesized structure implements the specified behavior. He need not simulate the design to compare the functionality against the ISPS or laboriously examine the different design representations line by line attempting to find similarities. SEESAW makes it easy for the designer to query the body of information provided by CORAL that describes the relationships between the different design representations.

9.2.2 Iterative Design Tools

A second area of application for the linking information provided by CORAL is in iterative design. In the past, the only feedback path in the Workbench was through the designer. The designer would have to analyze the Register-Transfer Level description produced by the Workbench, decide on what changes to make, and then determine what portion of the VT he needed to focus his changes on in order to obtain the desired result. By providing an explicit, machine readable statement of the relationships between different design representations, CORAL helps automate the feedback path to the previous design representation.

In a slightly different vein, some design decisions are best guided by information about design descriptions farther down in the hierarchy that have yet to be produced. In the past, heuristic estimates were substituted for the actual information which did not yet exist, or else the decision was made without any regard to the missing information. The links between design representations provided by CORAL help solve this problem. Now, information from the results of a previous pass of the design system can be fed back to the corresponding parts of an earlier design representation, providing real world numbers to help guide those decisions.

A package of iterative design tools has been implemented in the CORAL program. Three of these tools are outlined below.

Manual Rebinding. Manual rebinding is an example of the first kind of iterative design where changes to be made to the synthesized structure must be translated into commands in terms of the VT so they may be processed by EMUCS. The manual rebinding tool allows a designer to force structural operations that were performed by different functional units in a design produced by EMUCS to be performed in the same functional unit in the next iteration. Alternatively, the designer can force operations performed by a single functional unit to be split up over two or more different units in the next pass. Similar directives give control over the storage of values in registers. These kinds of commands may be useful when specific situations dictate a solution different from that derived from the cost tables used to guide EMUCS's design decisions.

The manual rebinding tool has access to all of the linking information provided by CORAL. In executing the command, the designer is prompted for the names of the modules he wishes to change. The tool then prompts for a target module for each of the operations (values) performed (stored) by those modules. The user may give any module name for any operation regardless of whether or not the module existed in the previous design. He can specify that the functionality of the specified modules be folded in to a single module, that each operation (value) be assigned to its own module, or some combination in between. The tool determines which VT operators (values) correspond to the structural operations (values) designated. This is necessary since EMUCS only accepts design descriptions at the VT level and directives to it must be in terms of VT operators and values. The rebinding tool generates a shell script to rerun EMUCS with directives included to bind the operations (values) as specified. The remainder of the design is redone automatically by EMUCS.

The rebinding tool can guard against combining operations that are scheduled for the same control step in to a single functional unit and can guard against sharing a register between values that are loaded at the same time. It does not, however, currently have the information needed to guard against sharing a register between values with overlapping lifetimes. The tool cannot force operations into separate control steps so they may be combined into a single functional unit because the scheduler does not accept such directives.

Area estimation. The partitioning tool described in Chapter 4 is concerned with partitioning a design into separate parts. One reason for doing so might be to break up the implementation of the design onto multiple chips. One kind of information that could be very useful in this task is the area required to implement any given part of the behavioral specification. This is a case of information from a not yet created representation being desired to guide a design decision. At present, area is not directly considered when making these decisions. CORAL, however, supplies linking information with the potential to make real world area information available to the partitioning tool. A prototype tool has been built in the CORAL package to demonstrate the technique.

The basic question for which we would like to provide an answer is "How much area will this part of the behavior take to implement?" The true answer depends on a variety of design decisions that have not yet been made. The area estimation tool can, however, answer the question "How much area *did* this part of the behavior take to implement the last time we implemented it?" While the answers to the two questions will not necessarily be the same, the answer to the second will often be close enough to provide a reasonable estimate of the answer to the first. The area estimation tool provides an answer to the second question by following the links from all the operators and values in the selected part of the behavioral specification to the modules used to perform and store them in the previous version of the structure. The areas of these modules are added and the total reported to the designer. Ultimately, this tool could be integrated directly into the Workbench where it could be used automatically by the partitioning tool.

At present, the Workbench lacks the capacity to generate chip layouts from the structural design representations, hence, actual area numbers are not available for the modules. Estimates of the area for each module are generated from a table that lists an area per bit cost for each function. The most expensive function performed by a module is used to determine its area.

Process Formation Estimation. Process formation estimation is similar to area estimation in that it attempts to answer a question about a design that has not yet been produced. Chapter 3 describes a process formation transform that may be applied to the VT. The transform splits off a portion of the behavioral specification into a separate process that will have its own data path and controller. One question that arises is "How much extra hardware will be generated given that this part of the behavior is split off into its own process?" As with area estimation, a prototype tool has been built that provides an estimate based on a previous implementation of the behavior.

The process formation estimator is also part of the CORAL package. It follows links from the operators and values in the selected portion of the VT and finds all the modules used to implement this part of the behavior in a previous run of the Workbench. It then traces links *back* from those modules to find all the parts of the behavior they are used to implement. Those modules that are used to implement both the selected part of the behavior and some other part will likely have to be duplicated in the two data paths that will result from process formation. Those modules that are used only to implement the selected part of the behavior will probably not be duplicated. The tool then reports the number and bit size of all the modules involved and those that are expected to be duplicated.

The answer provided by this tool is only an estimate. Interactions between operations in different parts of the behavior may cause EMUCS to combine operations and values in modules differently than it did when the behavior was only a single process. Even so, the estimates have proven fairly accurate for all types of modules except multiplexors and busses. The algorithm used is not appropriate for these types of modules for the following reason. Suppose a multiplexor has two input ports, one is used to handle values that happen to be part of the newly formed process and the other is used for values that are part of the rest of the behavior. When the behavior is split, the multiplexor that would otherwise be duplicated would end up being two, one input multiplexors. Since one input multiplexors are not very useful, they will be eliminated. A tool with access to more information about the relationships *within* a design representation than is available to CORAL would probably be able to handle multiplexors and busses correctly.

9.2.3 Other Applications

There is a long list of possible applications for the linking information provided by CORAL. Two more applications that have not yet been built are briefly described here.

Symbolic Debugger Interface. One tool that a production design system should certainly have is a simulator for the structural level designs synthesized by the design tools. One problem with using such a simulator, though, is the relative lack of familiarity with the structural design on the part of the system user. The user must have an intimate knowledge of the synthesized design in order to set break points and examine values and know what the results reported by the simulator mean. Time spent with something like the SEESAW tool described above would help the user gain this knowledge, but the learning process could still take inordinate amounts of time for a large design.

One solution is to provide a symbolic debugger style interface to the structural simulator. Such an interface would allow the user to set break points and examine values in terms of the ISPS specification which the user provided and presumably understands. The interface would use the linking information provided by CORAL to translate commands given in terms of ISPS carriers and line numbers into terms of registers and control steps. A symbolic debugger style interface should provide significant help to the designer trying to run simulations of automatically synthesized designs.

Formal Verification. Until the entire set of design tools has been proven correct, there will be room to doubt whether the automatically produced design does, in fact, implement the specified behavior. CORAL provides an explicit list of what the design tools *think* are corresponding values in the behavioral specification and the structural implementation. A formal verification system could be used to attempt to *prove* that those values are indeed equivalent. If the outputs of the two descriptions are provably identical given identical inputs, then the behaviors may be considered to be equivalent (without regard to any timing constraints that may be imposed on the behavior).

A formal verification system would work by starting at a point of known equivalence between the systems (e.g. the start state) and *symbolicly simulating* each description until the next point of equivalence is reached. The resulting expressions may be compared and, if proven equivalent, demonstrate the equivalence of the two design representations. Since the difficulty of proving expressions equivalent is known to be NP hard [Cory83] and the length of the expressions grows exponentially with the number of steps between comparisons, it is clear that it is advantageous to maximize the number of points of comparison. The detailed list of corresponding values supplied by CORAL helps minimize the work required to prove the implementation correct. In fact, a system based on CORAL's linking information could compare every equivalent value, including intermediate values within an ISPS expression or a Register-Transfer Level control state. It need not wait for the possibly infrequent points at which the states of the two descriptions are considered equivalent.

A formal verification system based on CORAL's linking information could raise confidence in the design tools and the designs produced and help in testing new design tools as they are added.

9.3 SUMMARY

By providing information about the relationships that exist between the initial and final representations in the Workbench, CORAL forms a basis for demonstrating the power provided by a synthesis system.

Knowledge about these relationships that exists within the design tools is captured and combined to tie together design descriptions in the different domains and different levels of abstraction. Thus, future tools tied into a synthesis system will have available both a high and low level representation of the design to work with. As discussed, there are a large number of design and analysis tools that can be built to make use of this information.

10. Observations and Future Work

While the research projects presented here combine to demonstrate Algorithmic and Register-Transfer Level Synthesis, they are far from the final solution. Rather they each have served to demonstrate basic concepts and approaches to the steps in synthesis, and together have shown two methods of integrating the steps into a synthesis system. In this Chapter, we make observations based on the results of the synthesis programs presented in the previous chapters and suggest future work in the area.

10.1. ARE THE TWO SYNTHESIS PATHS DIFFERENT?

At the beginning, we highlighted several differences in the approaches taken in the two paths through the System Architect's Workbench. These differences were: the use of style-specific vs. general synthesis algorithms, the handling of interactions between synthesis steps, and the method of knowledge representation.

From the results, we have seen that the style-specific approach wins in terms of design quality and runtime, but only when a design fits into a specific style or class of machine. A real issue in design is that most designs do not conform to only one style. Thus future synthesis programs must either improve on general synthesis algorithms or develop style-specific synthesis algorithms capable of first splitting up the design into its appropriate subparts, synthesizing each part with a style-specific algorithm, and finally combining the parts. In either case, this suggests that design tools such as APARTY will be important in identifying the different partitions that could be implemented using different styles.

Probably the biggest win for the style-specific approach was that it included a set of bus models from which to select. Since these models were typical of designs using that style, a starting plan for the design could be arrived at quickly and the problem of Register-Transfer Level design reduced to determining how to expand on the selected structure to implement the behavior. Beyond the selection step, many of the other techniques developed for the SUGAR system are applicable to a general approach to design. This suggests that a fruitful approach to developing general synthesis techniques may be to select from an expanded set of bus models based on algorithm analysis and technology capabilities and then use algorithms such as those developed in both of the Workbench's synthesis paths to complete the design. Essentially, the bus models would become the bases for different design styles.

The style-specific approach gained some advantage from knowing how different parts of the behavioral description mapped onto its notion of high level structure. For SUGAR, this was accomplished by annotating the description with qualifiers. As discussed previously, research on APARTY has shown how this partitioning can be done automatically.

The results have also highlighted two approaches to handling the interactions between the synthesis steps. The general synthesis path uses APARTY to suggest high level structure for the behavior and then uses this structure to guide the later synthesis steps. Assuming that APARTY can make informed decisions, then the later steps have a reduced search space to consider. Results have shown that APARTY can reduce the interconnect needed when CSTEP, EMUCS, and Busser use its guidance. Further, these four tools have been used with good results on a variety of designs spanning many styles. Coupled with the Seattle Silicon ChipCrafter System, a gate level simulation of part of the design using extracted circuit parameters demonstrated functional and timing correctness.

The style-specific SUGAR approach uses bus selection and delayed binding to control its interactions. Essentially the bus model selected suggests the high level structure, and the delayed binding splits the major synthesis steps (e.g. control sequencing) into fine-grain steps and then interleaves them. The delayed binding approach provides a more flexible synthesis approach in that it allows finer grained interactions to be considered.

The mixture of purely algorithmic and rule-based knowledge representation methods is an appropriate approach to follow. Where algorithmic solutions are available, they clearly win. However, the higher levels of design are difficult to quantify and thus rule-based approaches will be appropriate in the initial stages of CAD system development until more algorithmic solutions can be discovered.

In summary, this research has shown the importance of handling the interactions between the major Algorithmic and Register-Transfer Level synthesis steps. Good algorithms are known for solving each of the separate steps. The real problems now lie in making them work together. This research has reinforced the notion that it is necessary to guide the design based on high level analyses (e.g. design style selection, partitioning, bus selection), bottom-up information (e.g. technology cost tables), and fine grained interactions between the synthesis steps. Indeed, it would be interesting to mix the capabilities of the APARTY tool with a more general version of SUGAR's delayed binding techniques and some of CSTEP's, EMUCS', and Busser's algorithms.

10.2. YOU NEED MORE THAN SYNTHESIS

Too often, the synthesis researchers and commercial synthesis tool developers fail to realize that synthesis alone is not the end. Rather, it is a means to tie higher level abstractions and higher level designers into a comprehensive design system. The CORAL project laid the foundation for demonstrating this idea.

For instance, synthesis has often been rejected because the designer cannot easily understand what the synthesis tool has done. SEESAW remedies this problem by using the CORAL linked representation to display the resulting design in light of the original specification. As suggested in Chapter 9, further application of CORAL's linked representation promises advances in other design tasks because they now will have access to more appropriate information. Such advances will increase the usefulness of synthesis to the point that designers will demand its use.

10.3. ALGORITHMIC LEVEL SYNTHESIS

There is a growing belief, partially supported by the results presented here and those obtained by other synthesis research, that the Register-Transfer Level synthesis step is close to being solved and that future research in the area will only make minor contributions. Although there is some truth to that statement, namely that the area has become well-defined and reasonable solutions have been developed to the synthesis steps, current synthesis techniques are far from handling designs of the size available through present day technology. Thus, research in Register-Transfer Level synthesis will continue to be fruitful.

However, given that Register-Transfer Level synthesis is well-defined, it is time to expand our research efforts into the Algorithmic Level. The research presented here in behavioral transformations and architectural partitioning are only a start in this area. Major contributions addressing hardware/software partitioning, design style selection, and Algorithmic Level modeling will be made based on the availability of Register-Transfer Level synthesis techniques.

10.4. LOGIC SYNTHESIS, MODULE GENERATION AND PHYSICAL DESIGN

It is important to carry a synthesized design through physical design to completion for final evaluation. The primary concern is whether the interactions between the design decisions made at the high levels and those made in physical design restrict the quality of the design. As with the steps in Register-Transfer Level synthesis, either these interactions must be reduced through iteration in the design process so that convergence on a good design is obtained, or through good planning, physical design estimation, and top-down synthesis algorithms (that use bottom-up technology information) to follow the plan. Of course, there is room for a mixture of approaches.

Another controversy is whether Register-Transfer Level synthesis tools should be followed by logic synthesis tools, or by module generation tools. The approach taken by the Workbench so far has been to use module generation tools. The general synthesis path of the Workbench uses APARTY to develop an early notion of the structure of the algorithm to be implemented. This high level structure is further detailed by the Register-Transfer Level synthesis tools. The module generation tools tried to date have taken advantage of this structural hierarchy that extends above the logic module level. The untested thesis is that for data path oriented designs (as compared to control oriented designs), the module generation approach will create less physical design interactions with the high level synthesis tools. Considerable research is needed in this area.

10.5. DESIGN LANGUAGES

The research presented here has all been based on the use of the ISPS language as a behavioral input. Now that VHDL and Verilog have become available, we are developing compilers from these languages into the Workbench.

The behavioral semantics of all three languages are about the same, and thus the value trace format can be used with minor extensions for these languages. But since VHDL and Verilog also describe structural entities, they provide extra capabilities in mixed structure and behavior input. Such input is important in future synthesis systems because many digital systems are designed from previously designed hardware and new behavior. Thus languages such as these are preferable to the ISPS language for future synthesis research.

However, these languages were designed for simulation and their use for synthesis requires ignoring certain simulation-oriented assumptions. Among others, VHDL sequential behavior statements are assumed to execute in zero time! A key to the acceptance of Algorithmic and Register-Transfer Level synthesis will be the development of appropriate languages based on the needs of synthesis and simulation.

10.6. SUMMARY

Synthesis tools hold the promise of increasing productivity and expanding the horizon of IC applications to other engineering and science disciplines. It is clear from the results presented here that the designers of mass produced, leading edge digital systems such as single-chip processors would not use these or any other current synthesis tools. Indeed, high level design and IC technology are so dynamic that synthesis tools will always lag the leading-edge designer. However, these and other synthesis tools are becoming good enough to be used for the design of application specific integrated circuits. The first application will probably be with engineers and scientists who have an algorithm to put on a chip and a single chip microcomputer will not suffice. As synthesis capabilities increase, more designers will take advantage of their time saving features and the use of Algorithmic and Register-Transfer Level synthesis tools will expand greatly.

References

Note: Section numbers listing the places where references are used are listed after each reference.

[Agnew82] P.W. Agnew and A.S. Kellerman. Microprocessor Implementation of Mainframe Processors by Means of Architecture Partitioning. *IBM Journal of Research and Development* 26(4):401-412, July, 1982.

4.1.2, 4.6

[Aho86] A.V. Aho, R. Sethi and J.D. Ullman. *Compilers, Principles, Techniques, and Tools.* Addision-Wesley Publishing Company, 1985.

7.1.2

[Andrews83] Gary R. Andrews and Fred B. Schneider. Concepts and Notations for Concurrent Programming. *ACM Computing Surveys* 15(1):3-43, March, 1983.

3.3.1

[Barbacci73] M.R. Barbacci.
Automated Exploration of the Design Space for Register Transfer (RT) Systems.
PhD dissertation, Department of Computer Science, Carnegie-Mellon University, November 1973.

1.4

[Barbacci80] Mario R. Barbacci, Gary E. Barnes, Roderic G. Cattell, and Daniel P. Siewiorek.
The ISPS Computer Description Language.
Third edition, CS & EE Depts., Carnegie-Mellon University, 1980.

2.2.1

[Barbacci81] Mario R. Barbacci.
Instruction Set Processor Specifications (ISPS): The Notation and Its Applications.
IEEE Transactions on Computers C-30(1):24-40, January, 1981.

2.2.1

[Bell71] C.G. Bell and A. Newell.
Computer Structures: Readings and Examples.
McGraw-Hill Book Company, New York, 1971.

2.2.1

[Bhasker87] J. Bhasker and T. Samad.
A Better Clique-Partitioning Algorithm.
Internal report, Honeywell, Corporate Systems Development Division, 1987.

7.6.4

[Blackburn88a] Robert L. Blackburn, Donald E. Thomas, and Patti M. Koenig.
CORAL II: Linking Behavior and Structure in an IC Design System.
In *Proc. of the 25th DAC*, pages 529–535, June 1988.

9.1

[Blackburn88b] Robert L. Blackburn.
Relating Design Representations in an Automated IC Design System.
PhD Thesis, Carnegie Mellon University, October 1988.

9.1

[Borriello88] G. Borriello and E. Detjens.
High Level Synthesis: Current Status and Future Directions.
In *Proc. of the 25th DAC*, pages 477-482. ACM/IEEE, Anaheim, California, June, 1988.

8.1

[Brayton87] R.K. Brayton, R. Camposano, G. De Micheli, R.H.J.M. Otten, and J. van Eijndhoven.
The Yorktown Silicon Compiler System.
In *Silicon Compilation*, Daniel D. Gajski (Editor), pages 204–310, Addison-Wesley, 1988.

5.2.2

[Brewer86] F.D. Brewer and D.D. Gajski.
An Expert-System Paradigm for Design.
In *Proc. of the 23rd DAC*, pages 62-68. ACM/IEEE, Las Vegas, Nevada, June, 1986.

1.3.2

[Brewer87] Forrest D. Brewer and Daniel D. Gajski.
Knowledge Based Control in Micro-Architecture Design.
In *Proc. of the 24th DAC*, pages 203-209. ACM/IEEE, Miami, Florida, June, 1987.

1.3.1, 1.3.3, 6.1.5

[Campbell82] John E. Campbell and Joseph Tahmoush.
Design considerations for a VLSI Microprocessor

IBM Journal of Research and Development 26(4):454-463, July, 1982.

4.1.3

[Camposano86] Raul Camposano and Arno Kunzmann.
Considering Timing Constraints in Synthesis from a Behavioral Description.
In *Proc. of ICCD'86*, pages 6–9. Rye, New York, October 1986.

5.1.2

[Camposano87a] R. Camposano and R.K. Brayton.
Partitioning Before Logic Synthesis.
In *Proc. of ICCAD-87*, pages 324-326. IEEE, Santa Clara, California, November, 1987.

4.2

[Cattell82] R.G.G. Cattell.
Formalization and Automatic Derivation of Code Generators.
UMI Research Press, 1982.

7.4.2

[Cory83] W.E. Cory.
Verification of Hardware Design Correctness: Symbolic Execution Techniques and Criteria for Consistency.
PhD Thesis, Stanford University, June 1983.

9.2.3

[Davidson81] S. Davidson, D. Landskov and P.W. Mallett.
Some Experiments in Local Microcode Compaction for Horizontal Machines.
In *IEEE Design and Test* 3(6):460-477, July ,1981.

5.2.1

[De Man86] H. De Man, J. Rabaey, P. Six, and L. Claesen.
CATHEDRAL II: A Silicon Compiler for Digital Signal Processing.
In *IEEE Design and Test*, pages 13–25, December 1986.

1.3.2

[Dewilde85] P. Dewilde, E. Deprettere, and R. Nouta.
Parallel and Pipelined VLSI Implementation of Signal Processing Algorithms.
In S.Y. Kung, H.J. Whitehouse, and T. Kailath (editor), *VLSI and Modern Signal Processing*, chapter 15, pages 257-264. Prentice-Hall, 1985.

8.1

[Dickhut80] D. Dickhut, B. Hashizume and W.N. Johnson.
LSI Trio Calls the Tunes in Microcomputer's CPU.
Electronics 53:130-135, July 17, 1980.

4.1.3

[Dijkstra68] E.W. Dijkstra.
Co-operating Sequential Processes.
In Genuys (editor), *Programming Languages*, pages 43-112. Academic Press, 1968.

3.3.1

[Director81] Stephen W. Director, Alice C. Parker, Daniel P. Siewiorek, and Donald E. Thomas, Jr.
A Design Methodology and Computer Aids for Digital VLSI Systems.
In *IEEE Trans. on Circuits and Systems*, pages 634–635, July 1981.

1.4

[Druke80] M. Druke, et. al.
LSI Processor Mirrors High-Performance Minicomputer
Electronics :119-126, February 14, 1980.

4.1.3

[Dubois88] Michel Dubois, Christoph Scheurich, and Faye A. Briggs.
Synchronization, Coherence, and Event Ordering in Multiprocessors.
IEEE Computer, 21(2):9-21, February, 1988.

3.3.5

[Faggin78] F. Faggin.
How VLSI Impacts Computer Architecture.
IEEE Spectrum :28-31, May, 1978.

4.1.2

[Fisher79] J. A. Fisher.
The Optimization of Horizontal Microcode Within and Beyond Basic Blocks: An Application of Processor Scheduling with Resources.
PhD dissertation, Department of Mathematics and Computing, New York University, October 1979.

8.4.2

[Fisher81] J. A. Fisher.
Trace Scheduling: A Technique for Global Microcode Compaction.
IEEE Trans. on Computers, C-30(7): 478-490, July 1981.

5.3.1

[Gajski83] Daniel D. Gajski and Robert H. Kuhn.
Guest Editors' Introduction: New VLSI Tools.
IEEE Computer 16(12):11-14, December, 1983.

2.1, 2.1.5

[Gajski86] D. D. Gajski, N. D. Dutt, and B. M. Pangrle.
Silicon Compilation (Tutorial)
In *Proc. of Custom Integrated Circuits Conference*, pages 102-110. May 1986.

5

[Gatenby81] David A. Gatenby.
Digital Design from an Abstract Algorithmic Representation: Design and implementation of a Framework for Interactive Design.
Master's Thesis, EE Dept., Carnegie-Mellon University, October 1981.

5.2.1

[Girczyc84] E .F. Girczyc and J.P. Knight.
An Ada to Standard Cell Hardware Compiler Based on Graph Grammars and Scheduling.
In *Proc. of ICCD'84*, pages 726–731. Rye, New York, October 1984.

5.2.2

[Girczyc85] Emil F. Girczyc, Ray J.A. Buhr, and John P. Knight.
Applicability of a Subset of Ada as an Algorithmic Hardware Description Language for Graph-Based Hardware Compilation.
IEEE Trans. on CAD, pages 134–142, April 1985.

5.1.2

[Granacki85] John Granacki, David Knapp, and Alice Parker.
The ADAM Advanced Design Automation System: Overview, Planner and Natural Language Interface.
In *Proc. of the 22nd DAC*, pages 727-730. ACM/IEEE, Las Vegas, Nevada, June, 1985.

2.1.5

[Hitchcock83a] Charles Y. Hitchcock III and Donald E. Thomas.
A Method of Automatic Data Path Synthesis.
In *Proc. of the 20th DAC*, pages 484-489. ACM/IEEE, Miami, Florida, June, 1983.

1.3.1, 1.3.3, 6

[Hitchcock83b] Charles Y. Hitchcock III.
Automated Synthesis of Data Paths.
Master's Thesis, Carnegie Mellon University, January, 1983.

5.2.1, 6

[Hoare78] C.A.R. Hoare.
Communicating Sequential Processes.
CACM 21(8):666-677, August, 1978.

3.3.1

[I8086] Co-Processor Cooperates with 8 or 16-bit Microprocessors.
Electron. Design :19, March 1, 1980.

4.1.2

[Intel79] Intel Corporation.
Intel Multibus Interfacing: Application Note AP-28A.
Intel Corporation, Santa Clara, California, 1979.

5.4.1

[Johnson67] Stephen C. Johnson.
Hierarchical Clustering Schemes.
Psychometrika :241-254, September, 1967.

4.2

[Knapp84] David W. Knapp and Alice C. Parker.
A Data Structure for VLSI Synthesis and Verification.
Digital Integrated Systems Center Report DISC/83-6a, EE Dept.-Systems, University of Southern California, March, 1984.

2.1, 2.1.5

[Koenig87] Patti M. Koenig.
SEE-SAW: A Graphical Interface for System Level Design.

Master's Thesis, Dept. of Electrical and Computer Engineering, Carnegie Mellon University, December 1987.

9.2.1

[Kowalski84] Thaddeus Julius Kowalski.
The VLSI Design Automation Assistant: A Knowledge-Based Expert System.
PhD Thesis, Carnegie Mellon University, April 1984.

1.3.1, 1.3.2, 1.3.3, 4.2, 6.1.6

[Kurdahi87] Fadi J. Kurdahi and Alice C. Parker.
REAL: A Program for REgister ALlocation.
In *Proc. of the 24th DAC*, pages 210-215. ACM/IEEE, Miami, Florida, June, 1987.

6.1.4

[Lagnese89a] Elizabeth Dirkes Lagnese.
Architectural Partitioning for System Level Design of Integrated Circuits.
PhD Thesis, Dept. of Electrical and Computer Engineering, Carnegie Mellon University, March 1989.

4.4.7, 8.1.4, 8.1.6

[Lagnese89b] E. Dirkes Lagnese and D. E. Thomas.
Architectural Partitioning for System Level Design.
In *Proc. of the 26th DAC*, pages 62-67. ACM/IEEE, Las Vegas, Nevada, June, 1989.

8.1.4, 8.1.6

[Leighton79] F.T. Leighton.
A Graph Coloring Algorithm for Large Scheduling Problems.
In *Jour. of Research of the National Bureau of Standards* 84(6):489-506, November-December 1979.

7.6.1

[Leverett80] B.W. Leverett, R.G.G. Cattell, S.O. Hobbs, J.M. Newcomer, A.H. Reiner, B.R. Schatz, and W.A. Wulf. An Overview of the Production-Quality Compiler-Compiler Project. In *Computer* 13(8):38-49, August 1980.

7.2.1

[Marwedel86] Peter Marwedel. A New Synthesis Algorithm for the MIMOLA Software System. In Proc. *of the 23rd DAC*, pages 271–277, June 1986. ACM/IEEE, Las Vegas, Nevada, June, 1986.

1.3.3, 6.1.2

[McFarland83] M.C. McFarland S.J. Computer-Aided Partitioning of Behavioral Hardware Description. In *Proceedings of the 20th DAC*, pages 472-476. ACM/IEEE, June, 1983.

4.2

[McFarland86a] Michael C. McFarland, S.J. Using Bottom-Up Design Techniques in the Synthesis of Digital Hardware from Abstract Behavioral Descriptions. In *Proc. of the 23rd DAC*, pages 474-480. ACM/IEEE, Las Vegas, Nevada, June, 1986.

1.3.1, 1.3.3, 4.2,6.1.6

[McFarland86b] Michael C. McFarland, S.J. and Thaddeus J. Kowalski. Assisting DAA: The Use of Global Analysis in an Expert System. In *Proc. of ICCD'86*, pages 482–485. Rye, New York, October 1986.

5.2.1

[McFarland88] Michael C. McFarland, S.J., Alice C. Parker, and Raul Camposano.
Tutorial on High-Level Synthesis.
In *Proceedings of the 25th DAC*, pages 330-336. ACM/IEEE, Anaheim, California, June 1988.

1.1.2, 6.1

[Murgai89] Rajeev Murgai.
Automatic Design of Control Paths for System Level Synthesis.
Master's Thesis, Dept. of Electrical and Computer Engineering, Carnegie Mellon University, January 1989.

8.2.4

[Nestor87] John Anthony Nestor
Specification and Synthesis of Digital Systems with Interfaces.
PhD Thesis, Carnegie Mellon University, April, 1987.

5.1.2, 5.4.1, 6

[Oakley79] J. Oakley.
Symbolic Execution of Formal Machine Descriptions.
PhD dissertation, Department of Computer Science, Carnegie-Mellon University, 1979.

7.2.2

[Park88] N. Park and A. C. Parker.
Sehwa: A Software Package for Synthesis of Pipelines from Behavioral Specifications.
IEEE Trans. on CAD, pages 356-370, March, 1988.

3.6

[Parker79] Alice C. Parker, Donald E. Thomas, Stephen Crocker, and Roderic G.G. Cattell.
ISPS: A Retrospective View .

In *Proc. 4th Int. Symp. on CHDLs*, pages 21-27. ACM/IEEE, Palo Alto, California, October, 1979.

2.2.1, 3.3.2

[Parker86] A.C. Parker, J. Pizarro, and M. Milnar. MAHA: A Program for Data Path Synthesis. In *Proceedings of the 23rd DAC*, pages 461-466. ACM/IEEE, Las Vegas, Nevada, June, 1986.

1.3.1,.1.3.3, 5.1.2, 5.2.2, 6.1.4

[Paulin86] P.G. Paulin, J.P. Knight, and E.F. Girczyc. HAL: A Multi-Paradigm Approach to Automatic Data Path Synthesis. In *Proc. of the 23rd DAC*, pages 263–270. ACM/IEEE, Las Vegas, Nevada, June, 1986.

[Paulin87] P.G. Paulin and J.P. Knight. Force-Directed Scheduling in Automatic Data Path Synthesis. In *Proc. of the 24th DAC*, pages 195-202. ACM/IEEE, Miami, Florida, June, 1987.

1.3.3, 5.2.2, 6.1.1

[Peng86] Zebo Peng. Synthesis of VLSI Systems with the CAMAD Design Aid. In *Proc. of the 23rd DAC*, pages 278-284. ACM/IEEE, Las Vegas, Nevada, June 1986.

6.1.3

[Pangrle87] Barry Michael Pangrle and Daniel D. Gajski. Design Tools for Intelligent Silicon Compilation. *IEEE Trans. on CAD*, pages 1098–1112, November 1987.

5.1, 5.2.1

[Principles80] *IBM System/370 Principles of Operation.* Seventh Edition, March 1980.

4.6

[Rabaey88] J. Rabaey, H. De Man, J. Vanhoof, G. Goossens, and F. Catthoor.
CATHEDRAL-II: A Synthesis System for Multiprocessor DSP Systems.
In *Silicon Compilation*, Daniel D. Gajski (Editor), pages 311-360, Addison-Wesley, 1988.

[Rajan85] Jayanth V. Rajan and Donald E. Thomas.
Synthesis By Delayed Binding of Decisions.
In *Proceedings of the 22nd DAC*, pages 367-373. ACM/IEEE, 1985.

4.2

[Rajan88] Jayanth V. Rajan.
Automatic Synthesis of Microprocessors.
PhD Thesis, Dept. of Electrical and Computer Engineering, Carnegie Mellon University, December 1988.

7.5, 7.7, 8.5

[Schmult84] Brian Schmult.
Partitioning Strategies for Muli-Chip VLSI Microprocessors.
Master's Thesis, Carnegie Mellon University, February, 1984.

4.1.2

[Snow78] Edward A. Snow.
Automation of Module Set Independent Register-Transfer Level Design.
PhD thesis, EE Dept., Carnegie-Mellon University, April, 1978.

2.2, 2.2.2, 7.2.2

[Thomas77] Donald E. Thomas.
The Design and Analysis of an Automated Design Style

Selector.
PhD Thesis, Dept. of Electrical Engineering, Carnegie Mellon University, March 1977.

1.3.2

[Thomas83] D.E. Thomas, C.Y. Hitchcock, III, T.J. Kowalski, J.V. Rajan, and R.A. Walker.
Methods of Automatic Data Path Synthesis.
IEEE Computer 16(12):59-70, December, 1983.

1.4, 5.2.1, 8.4.3

[Thomas88] D.E. Thomas, E.M. Dirkes, R.A. Walker, J.V. Rajan, J.A. Nestor, and R.L. Blackburn.
The System Architect's Workbench.
In *Proc. of the 25th DAC*, pages 337–343. ACM/IEEE, Anaheim, California, June, 1988.

1.3.2, 1.4

[Tredennick80] N. Tredennick.
Implementation Decisions for the MC68000 Microprocessor.
In *Proc. of the 3rd Rocky Mountain Symp. on Microcomputers: Systems, Software, Architecture*, 1980, pages 87-102.

3.2.4

[Trickey87] Howard Trickey.
Flamel: A High-Level Hardware Compiler.
In *IEEE Trans. on CAD* 6(2):259–269, March 1987.

1.3.1

[Trick89] Michael T. Trick and Stephen W. Director.
LASSIE: Structure to Layout for Behavioral Synthesis Tools.
In *Proc. of the 26th DAC*, pages 104-109. ACM/IEEE, Las Vegas, Nevada, June, 1989.

8

[Tseng83] Chia-Jeng Tseng and Daniel P. Siewiorek.
Facet: A Procedure for the Automated Synthesis of Digital Systems.
In *Proc. of the 20th DAC*, pages 490–496, June 1983.

1.3.1, 1.3.3

[Tseng84a] Chia-Jeng Tseng.
Automated Synthesis of Data Paths in Digital Systems.
PhD Thesis, Carnegie Mellon University, April, 1984.

5.2.1

[Tseng84b] Chia-Jeng Tseng.
Automated Synthesis of Data Paths in Digital Systems.
PhD Thesis, Dept. of Electrical and Computer Engineering, Carnegie Mellon University, April 1984.

8.4.3

[Tseng86] Chia-Jeng Tseng and Daniel P. Siewiorek.
Automated Synthesis of Data Paths in Digital Systems.
IEEE Trans. on CAD, pages 379–395, July 1986.

6.1.7, 6.6.1

[Tseng88] Chia-Jeng Tseng, Ruey-Sing Wei, Steven G. Rothweiler, Michael M. Tong, and Ajoy K. Bose.
Bridge: A Versatile Behavioral Synthesis System.
In *Proc. of the 25th DAC*, pages 415–420. ACM/IEEE, Anaheim, California, June 1988.

5.1

[Un89] Edward D. Un.
Evaluating Register-Transfer Level Synthesis Tools Using Gate Level Simulation.
Master's Thesis, Dept. of Electrical and Computer Engineering, Carnegie Mellon University, January 1989.

8.2.4

[Vasantharajan82] Jayanth Vasantharajan.
Design and Implementation of a VT-Based Multi-Level Representation.
Master's thesis, EE Dept., Carnegie-Mellon University, February, 1982.

3.8

[Vegdahl82] S.R. Vegdahl.
Local Code Generation and Compaction in Optimizing Microcode Compilers.
PhD dissertation, Department of Computer Science, Carnegie-Mellon University, December 1982.

7.4

[Walker83] Robert A. Walker and Donald E. Thomas.
Behavioral Level Transformation in the CMU-DA System.
In *Proc. of the 20th DAC*, pages 788-789. ACM/IEEE, Miami, Florida, June, 1983.

7.2.2

[Walker87] R.A. Walker and D.E. Thomas.
Design Representation and Transformation in The System Architect's Workbench.
In *Proc. of ICCAD-87*, pages 166-169. IEEE, Santa Clara, California, November, 1987.

3

[Walker88] Robert A. Walker
Design Representation and Behavioral Transformation for Algorithmic Level Integrated Circuit Design.
PhD Thesis, Carnegie Mellon University, April 1988.

3, 8.1.1, 8.1.4

Index